Architectural Press Legal Guides

Dr David Chappell RIBA MA PhD currently lectu[...] contractual procedures. He has previously work[...] private-sector practice and has experience as con[...] contractor. He is author of *Contractual Corresponde[...]* *Architects*, and *Contractor's Claims*. He is also joint author with Dr Vincent Powell-Smith of *Building Contract Dictionary*.

Dr Vincent Powell-Smith LLM DLitt FCIArb, Commander of the Order of Polonia Restituta, sometime Lecturer in Law at the University of Aston Management Centre, now acts as a consultant specialising in building contracts and as a practising arbitrator. A well known conference speaker, he has been Legal Correspondent of *Contract Journal* for the past twelve years and is a regular contributor to *The Architects' Journal* and *International Construction*. A former member of the Council of the Chartered Institute of Arbitrators, for several years he was a member of the Minister's Joint Advisory Committee on Health and Safety in the Construction Industry. He has written a number of highly successful titles on construction law topics.

Architectural Press Legal Guides

JCT Intermediate Form of Contract
An Architect's Guide

DAVID CHAPPELL and VINCENT POWELL-SMITH

The Architectural Press: London

First published in 1985 by the Architectural Press Ltd,
9 Queen Anne's Gate, London SW1H 9BY

© D M Chappell and V Powell-Smith 1985

All rights reserved. No part of this book may be reproduced, stored in a retrieval system or transmitted in any form or by any means, electronic, mechanical, photocopying or otherwise, for publication purposes, without the prior permission of the publisher. Such permission, if granted, is subject to a fee, depending on the nature of the use.

The only exceptions to the above prohibition are the sample letters provided in this book as models to be used by readers in architectural practice. These letters may be freely reproduced, on the clear understanding that the publishers and the author do not guarantee the suitability of any particular letter for any particular situation. Legal advice should be taken in case of doubt.

British Library Cataloguing in Publication Data
Chappell, David
    JCT intermediate form of contract: an architect's
guide.—(Architectural Press legal guides; 3)
    1. Joint Contracts Tribunal—Intermediate form
of building contract. 1984   2. Building—
Contracts and specifications—Great Britain
    I. Title   II. Powell-Smith, Vincent
    692'.8   TH425

ISBN 0–85139–885–5

Typeset by Phoenix Photosetting, Chatham
Printed in Great Britain by Dotesios Printers Ltd, Bradford-on-Avon

Contents

	Preface	xi
1	**The purpose and use of IFC84**	1
1.1	The background	1
1.2	IFC documentation	2
1.3	The use of IFC84	3
1.4	Completing the form	4
2	**Contracts compared**	9
3	**Contract documents and insurance**	17
3.1	Contract documents	17
3.1.1	Types and uses	17
3.1.2	Importance and priority	19
3.1.3	Errors	21
3.1.4	Custody and copies	22
3.1.5	Limits to use	22
3.2	Insurance	23
3.2.1	Indemnity	23
3.2.2	Injury to persons and property	24
3.2.3	Things which are the liability of the employer	27
3.2.4	Clause 6.3 perils	30
3.2.5	A new building where the contractor is required to insure	30

3.2.6	A new building at the sole risk of the employer	31
3.2.7	Alterations or extensions to an existing building	31
3.3	Summary	32
4	**The architect's authority and duties**	**34**
4.1	Authority	34
4.1.1	General	34
4.1.2	Express provisions	41
4.1.3	The issue of instructions	43
4.1.4	Instructions in detail	48
4.2	Duties	58
4.2.1	Duties under the contract	58
4.2.2	General duties	59
4.3	Summary	60
5	**The contractor's obligations**	**61**
5.1	Express and implied obligations	61
5.1.1	Legal principles	61
5.1.2	Execution of the works	62
5.1.3	Workmanship and materials	71
5.1.4	Statutory obligations	72
5.1.5	Person-in-charge	72
5.1.6	Levels and setting out	73
5.2	Other obligations	73
5.2.1	Access to the works and premises	73
5.2.2	Drawings, details, and information	73
5.2.3	Obedience to architect's instructions	74
5.2.4	Other rights and obligations	76
5.3	Summary	76
6	**The employer's powers, duties and rights**	**77**
6.1	Express and implied powers and duties	77
6.1.1	Co-operation or non-interference	77
6.2	Rights	82

Contents

6.2.1	General	82
6.2.2	Deferment of possession of the site	82
6.2.3	Deduction/repayment of liquidated damages	83
6.2.4	Employment of direct contractors	86
6.2.5	Rights as to insurance	87
6.3	Duties	87
6.3.1	General	87
6.3.2	Payment	88
6.3.3	Retention	89
6.3.4	Other duties	89
6.4	Summary	89
7	**The clerk of works**	90
7.1	Appointment	90
7.2	Duties	91
7.3	Responsibility	92
7.4	Summary	93
8	**Subcontractors and suppliers**	94
8.1	General	94
8.2	Subcontractors	94
8.2.1	Assignment and subcontracting	94
8.2.2	Named persons as subcontractors	96
8.3	Statutory authorities	103
8.4	Work not forming part of the contract	104
8.5	Summary	106
9	**Possession, practical completion, and defects liability**	109
9.1	Possession	109
9.1.1	General	109
9.1.2	Date for possession	110
9.2	Practical completion	112
9.2.1	Definition	112
9.2.2	Consequences	113
9.3	Defects liability period	113
9.3.1	Definition	113
9.3.2	Defects, shrinkages, or other faults	115

Contents

9.3.3	Frost	116
9.3.4	Procedure	116
9.4	Summary	121
10	**Claims**	**122**
10.1	General	122
10.2	Extension of time	125
10.2.1	Legal principles	125
10.2.2	Liquidated damages	125
10.2.3	Procedure	127
10.2.4	Grounds	139
10.3	Loss and expense claims	143
10.3.1	Definition	143
10.3.2	Procedure	144
10.3.3	Matters grounding a claim	145
10.4	Summary	150
11	**Payment**	**152**
11.1	The contract sum	152
11.2	Payment before practical completion	154
11.2.1	Method and timing	154
11.2.2	Valuation	154
11.2.3	Amounts included	156
11.3	Payment at practical completion	158
11.4	Retention	158
11.5	Final payment	160
11.6	The effect of certificates	160
11.7	Variations	161
11.8	Fluctuations	168
11.9	Summary	169
12	**Determination**	**172**
12.1	Determination by the employer	172
12.1.1	General	172
12.1.2	Grounds (clause 7.1)	173
12.1.3	Grounds (clause 7.2)	180
12.1.4	Grounds (clause 7.3)	182
12.1.5	Grounds (clause 7.8.1)	182

Contents

12.1.6	Grounds (clause 6.3C.2)	183
12.1.7	Consequences (clause 7.4)	185
12.1.8	Consequences (clauses 7.9 and 6.3C.2)	188
12.2	Determination by the contractor	188
12.2.1	General	188
12.2.2	Grounds (clause 7.5)	189
12.2.3	Grounds (clause 7.6)	195
12.2.4	Grounds (clause 7.8.1)	196
12.2.5	Grounds (clause 6.3C.2)	196
12.2.6	Consequences (clause 7.7)	196
12.2.7	Consequences (clauses 7.9 and 6.3C.2)	198
12.3	Summary	199
13	**Arbitration**	200
13.1	General	200
13.2	Appointing an arbitrator	203
13.3	An arbitrator's powers	205
13.4	Appeals to the High Court	207
13.5	Third-party procedure	210
13.5	Summary	211
Appendix A	**Form of Tender and Agreement NAM/T**	212
Appendix B	**NAM/SC Subcontract conditions**	216
Appendix C	**The RIBA/CASEC Form of Employer/Specialist Agreement ESA/1**	226
Appendix D	**Clause number index**	230

Preface

Ever since the 1980 edition of the Standard Form of Building Contract (JCT80) was published by the Joint Contracts Tribunal, there has been a demand for a form of contract which was less complicated than the full standard form, but more comprehensive than the Agreement for Minor Building Works 1980 (MW80). After much drafting and redrafting, the JCT Intermediate Form of Building Contract (IFC84) was published in September 1984 to satisfy that demand.

How far the latest of the JCT forms fulfils expectations will depend on the individual. The structure follows that of MW80, but in detail it owes much to JCT80. It appears to be shorter than it really is, because it is laid out in two columns. The language is no less complex than that of JCT80 and, in places, there are ambiguities. Nonetheless, the form will probably find favour for the vast number of medium-sized contracts for which MW80 is clearly inadequate.

Although this book is addressed to the architect, as the person charged with the administration of the contract, it will be useful to contractors also, because it explains not only what the clauses mean but also their implications in practice.

The book is not a clause–by–clause interpretation. Instead, the form has been dealt with by considering the roles of the parties involved and by devoting separate chapters to important topics such as claims, payment, determination, and arbitration. Legal language has been avoided, and a series of flowcharts, tables, and sample letters has been included to make up a practical working tool.

The copyright in IFC84, Tender and Agreement NAM/T, and Employer/Specialist Agreement ESA/1, and in the supporting practice notes, is vested in RIBA Publications Ltd, and we are indebted to them for permission to quote from those publications for the purpose of this book.

David Chappell
Vincent Powell-Smith

1 The purpose and use of IFC84

1.1 The background

The JCT Intermediate Form of Building Contract (IFC84) was published in September 1984 and is the latest member of the growing family of standard contract forms issued by the Joint Contracts Tribunal. Table 1.1 sets out the JCT forms currently available.

It is intended that IFC84 should fill the gap between MW80, which is suited to smaller projects, and the very complex JCT80 with its lengthy and complicated provisions for nominated subcontractors. Its most significant features are as follows:

— the provisions (clause 3.3.) enabling the architect to select specialist subcontractors by 'naming' them
— the power for the employer to defer giving possession of the site for a period up to six weeks (clause 2.2.)
— an optional provision in the arbitration agreement, allowing points of law to be settled by reference to the High Court
— the absence of restraint on opening arbitration proceedings during the progress of the works

In arrangement, the form follows the pattern of MW80 in setting out the conditions under eight main headings, the clauses being printed in two columns, but the wording follows very closely that of JCT80 in many places.

IFC84 is arranged as follows:

ARTICLES OF AGREEMENT

CONDITIONS
1 *Intentions of the parties*
2 *Possession and completion*

The purpose and use of IFC84

3 *Control of the works*
4 *Payment*
5 *Statutory obligations, etc*
6 *Injury, damage, and insurance*
7 *Determination*
8 *Interpretation, etc*

APPENDIX

SUPPLEMENTAL CONDITIONS
A Value Added Tax
B Statutory tax-deduction scheme
C Contributions, levy, and tax fluctuations (published separately)
D Use of price adjustment formulae (published separately)
E Fair wages clause

The subdivisions within the main conditions are not always logical, and in this book we have adopted what we believe to be a more realistic arrangement.

1.2 IFC Documentation

In addition to the printed contract form itself, which you will need to complete and to arrange to have executed formally by the employer and the contractor, there is also a set of supporting documentation, as follows:
— Fluctuations Clauses and Formula Rules (Supp/IFC84)
— Tender and Agreement for Named Subcontractors under IFC84 (NAM/T)—summarised in Appendix A, p 212
— Subcontract Conditions for Named Subcontractors under IFC84 (NAM/SC)— summarised in Appendix B, p 216
— Subcontract Formula Rules for Named Subcontractors under IFC84 (NAM/SC/FR)

The Joint Contracts Tribunal has also published two related practice notes.
— Practice Note IN/1: Introductory notes on the JCT Intermediate Form of Building Contract IFC84
— Practice Note 20: Deciding on the Appropriate Form of JCT Main Contract—Revised July 1984

Independently, but in consultation with the JCT, the RIBA and CASEC have prepared a form of Employer/Specialist Agreement under IFC84 (ESA/1). This is summarised in Appendix C (p 226). It is a vital document. You will be responsible for arranging for its completion on the employer's behalf. Its effect is to create a direct contractual relationship between the employer and the named subcontractor for limited purposes, and it gives the employer the necessary protection against design and related failures in a named subcontractor's work.

1.3 The use of IFC84

IFC84 is issued for contracts in the range between JCT80 with quantities and MW80. It is suitable for use where the proposed works are
–   simple in content, involving the normally recognised basic trades and skills of the industry;
–   without any building service installations of a complex nature; and
–   adequately specified, or specified and billed as appropriate before the issue of tenders.

IFC84 is suitable for use only where these criteria are met. Practice Note 20 contains a little further guidance, from which it may be deduced that IFC84 is suitable for most such projects, with or without quantities, whose value is between £55,000 and £250,000 (at 1984 prices), with a contract period of not more than 12 months.

Contract value is not, however, the deciding factor, and the practice note points out that IFC84 may 'be suitable for somewhat larger or longer contracts' provided that the three basic criteria are met. Certainly, since IFC84 contains adequate provisions for contractors to recover fluctuations, the contract period is not the decisive factor.

Three things would preclude the use of the form:
–   complexity of work,
–   the wish or need for nominated subcontractors and nominated suppliers, or
–   the wish to make the contractor responsible for design, wholly or in part.

In summary, IFC84 is highly suitable for use on projects of medium size with a simple work content. You should not use IFC84 merely because you wish to avoid the complexities of nomination under JCT80, because in their own way the IFC provisions for named subcontractors are just as complicated!

Table 1.2 sets out the main differences between the Association of Consultant Architects' Form of Building Agreement 1984 (ACA 2), JCT80, and IFC84 to assist you to choose the most appropriate form.

The purpose and use of IFC84

1.4 Completing the form

Normally, a formal contract will be executed by the parties, and the printed form IFC84 will be used for this purpose.

The articles of agreement

Page 1 should be completed with the descriptions and addresses of the employer and the contractor. The date will not be inserted until the form is signed or sealed by the parties.

The first recital must be completed with special care because the description of 'the Works' is important in many respects, not least when considering the question of variations, and the appropriate deletions should be made to both the first and second recitals as indicated in the footnotes. Consistency is important, since the contractor is required to carry out and complete the works in accordance with the contract documents identified in the second recital. This is in two mutually exclusive alternatives, as described in Chapter 3.

Completion of articles 1 to 4 is routine and self–explanatory. Article 5—settlement of disputes by arbitration—requires special consideration in consultation with the employer, who must make the decision as to whether or not he wishes questions of law to be referred to the High Court. Probably, most private-sector employers will not choose this option, which is in any event available—should the need arise—under the general law (see Chapter 13).

Attestation

Alternative attestation clauses are provided, as the contract may be executed by hand or under seal. The employer will have made this decision—on your advice—at pre-tender stage, and there is an important practical difference.

The Limitation Act 1980 specifies a limitation period—the time within which an action may be commenced—of six years where the contract is merely signed by the parties or of 12 years in the case of a contract entered into under seal. These periods begin to run 'when the cause of action accrues', which, in a case of a contractual claim, is the date when the defective work was done (ie, the date of the breach of contract). It is usually sensible to contract under seal.

To do this, red wafer seals should be affixed, though it is sufficient to write 'L.S.' (meaning 'in the place of the seal') if wafer seals are not obtainable. It is then your responsibility to ensure that the contracts are stamped with an impressed 50p Inland Revenue stamp within 30 days of the contract being executed by the parties.

Both copies of the form, as sealed by the parties, should be sent by

1.4 Completing

post with 50p per copy to the Inland Revenue Postal Section, Adjudication Department, West Block, Barrington Road, Worthing, Sussex.

If — as is not generally desirable — any amendments are made to the printed text or any clauses are deleted, these should be initialled by both contracting parties.

The appendix

This should be fully and carefully completed, consistently with the information which you gave to the contractor at tender stage since his tender is based on the information supplied.

Appendix entries which are inconsistent with the printed conditions generate much income for lawyers!

Table 1.1
JCT forms of contract in current use

Title	Contract documents	Comments
Standard Form of Building Contract 1980 (JCT80)	Drawings Bills of quantities (*Variants*: with Approximate quantities *or* Specification)	Complex and sophisticated: for use on major projects where it is wished to use nominated subcontractors or suppliers
Intermediate Form of Building Contract (IFC84)	Drawings and *either* Specification *or* Schedules of work *or* Bills of quantities	Only suitable where contract is not long or complex and any specialist work is of a simple nature. Value range from £55,000 to £250,000 (1984 prices). Cannot be used where sectional completion is desired.
Agreement for Minor Building Works 1980 (MW80)	Drawings *and/or* Specifications *and/or* Schedules	Normal maximum value of £50,000 (1981 prices). Short contract period, as no provision for full labour and materials fluctuations, and only where simple conditions are appropriate
Fixed Fee Form of Prime Cost Contract	Specifications *with or without* drawings	Little used in practice—prime cost plus fixed fee
Standard Form with Contractor's Design	Employer's requirements Contractor's proposals Contract sum anaylsis	For use where contractor is to be responsible for all of the design. If the contractor is to be responsible for part of the design, JCT80 with quantities should be used together with the Contractor's Designed Portion Supplement which modifies the form.

The purpose and use of IFC84

Table 1.2 ACA2, JCT80, and IFC84 compared.

Subject	ACA2	JCT80	IFC84
Cost limits (upper)	None	£250,000+ but can be used for smaller projects with specialist work content or complex	£55,000 to £250,000—and works must be of simple content and with no specialist service
Contract documents	Drawings Schedule of rates *or* Bills of quantities Specification	Drawings Bills of quantities (Variants with approx. quantities or specification)	Drawings Bills of quantities *or* schedules of rates *or* specification
Date for possession	To be given to contractor as specified in time schedule Provision for possession to be given in parts	Must be given on date stated in appendix No power to postpone possession	As JCT80, but power to defer giving of possession for up to six weeks in return for time and money
Extension of time	Alternative clauses—11.5.1 covers 'any act, instruction, default or omission of employer or architect on his behalf'; The other similar to JCT80. Detailed provisions for notification	Detailed provisions and long list of grounds, not comprehensive and exact scope not clear	Similar listing to JCT80, but some omissions, and deferment of possession an optional ground Notification procedures less detailed
Liquidated damages	Alternative clauses—conventional liquidated damages *or* unliquidated damages. Architect's certificate of delay a precondition to deduction and architect empowered to make deduction. Interest to contractor if extension given later	Liquidated damages deductible by employer if architect certifies late completion and employer so requires. Provision for adjustment if new completion date fixed later but no provision for interest	Similar to JCT80

1.4 Completing

Variations	Detailed provisions—contractor can be required to submit estimate of cost, time, and loss and/or expense. Detailed valuation rules	Detailed provisions with valuation rules	Similar to JCT80
Payment and retention	Interim payments monthly but conditional on contractor making application vouchered 95%. Final account must be submitted by contractor within 60 days of expiry of maintenance period	Interim payments monthly unless otherwise stated. Final account	Broadly similar to JCT80—95% of value of work and materials in monthly interim certificates, 97% at practical completion
Fluctuations	Optional fluctuations clause based on ACA Index—80% only payable	Alternative provisions for contributions, levy, and tax fluctuations, or labour, materials and tax fluctuations, or formula adjustment where there are contract bills	As JCT80 except no option for labour and materials and tax fluctuations
Loss and/or expense	Detailed provisions entitling contractor to claim for disturbance of regular progress caused by employer's or architect's 'acts, omissions, default, of negligence'. Notice of claim required if payment to be included in interim certificate and estimates of cost required. Otherwise architect adjusts in Final Certificate but contractor loses by interest element	A long list of matters and detailed claims procedures; not comprehensive, and many claims must be dealt with at common law	Similar to JCT80 but deferment of possession also a ground

1.4 Completing

Table 1.2
ACA2, JCT80, and IFC84 compared—*cont.*

Subject	ACA2	JCT80	IFC84
Selection of subcontractors	Provision for named subcontractors and suppliers Architect can be involved in negotiations Contractor fully responsible for performance, including design failure	List of three 'selected' Detailed and bureaucratic provisions for nominated subcontractor and complex supporting documentation Contractor exempted from design responsibility Provision for nominated suppliers	Subcontractors can be named in contract documents or ps instruction No provision for naming suppliers
Disputes settlement	Alternative methods: — Conventional arbitration — Adjudication followed by arbitration — Litigation If adjudication option chosen, an immediate decision can be given which is binding on both parties unless upturned in arbitration	Conventional arbitration, generally cannot be opened while works still in progress	Similar to JCT80 but no restraint on arbitration while works in progress Optional provision for legal points to be settled by High Court
Advantages/ disadvantages	Flexible Simple payment scheme Provision for design responsibility by contractor with indemnity cover Useful range of alternative clauses Time periods not always realistic	Comprehensive Widely known Complex nominated subcontractor provisions Unrealistic fluctuations	Provision for deferment of possession Wide testing provisions Not suitable for long or complex contracts

2 Contracts compared

IFC84 is a very flexible contract, and its content establishes it as a member of the Joint Contracts Tribunal family. Its provisions closely parallel those of JCT80 in many respects, although the format and layout follow MW80. Like the latter, it is silent on many important matters: eg, there is no express provision which entitles the architect to access to the contractor's workshops, etc. Such gaps would have to be filled in by the common law.

The revised Practice Note 20 (July 1984), Deciding on the Appropriate Form of JCT Main Contract, contains some broad general guidance. Among other things, it compares the JCT contracts by reference to contract documents and price, and an appendix summarises the main differences in the contract conditions in matters of major significance in building contracts.

Table 2.1 compares IFC84 with JCT80 and MW80, and also with the second edition of the ACA Form of Building Agreement, published in September 1984 (ACA2).

Contracts compared

Table 2.1
IFC Clauses compared with those of other common standard contracts

IFC clause	Description	JCT80 clause	MW80 clause	ACA2 clause	Comment (on IFC84 unless otherwise stated)
1	**Intentions of the parties**				
1.1	Contractor's obligations	2	1.1	1.1	
1.2	Quality and quantity of work	14	—	—	
1.3	Priority of contract documents	2.2.1	—	1.3	The printed conditions, etc, prevail over any specially prepared clauses insofar as there is any conflict
1.4	Instructions as to inconsistencies, errors, or omissions	2.3 2.2.2.2	4.1	1.4 1.5	
1.5	Contract bills and SMM	2.2.2	—	1.4	
1.6	Custody and copies of contract documents	5.1	—	—	
1.7	Further drawings and details	5.4	1.2	2.1	
1.8	Limits to use of documents	5.7	—	—	
1.9	Issue of certificates by architect	5.8	1.2	23.1	
1.10	Unfixed materials or goods: passing of property, etc	16.1	—	6.1	These provisions would not be effective against a retention of title clause in a supplier's contract of sale

Contracts compared

1.11	Off-site materials and goods: passing of property, etc	16.2	—	6.1	These provisions would not be effective against a retention of title clause in a supplier's contract of sale
2	**Possession and completion**				
2.1	Possession and completion dates	23.1	2.1 2.4	11.1	
2.2	Deferment of possession				The power to defer the giving of possession is one unique to IFC84, but ACA clause 11.8 empowers architect to order acceleration and postponement
2.3	Extension of time	25	2.2	11.6	
2.4	Events referred to in 2.3	25.4	2.2	11.5	
2.5 2.6	Further delay or extension of time	25.1	—	11.6	
2.6	Certificate of non-completion	24.1	—	11.2	There is express power to issue further certificates of delay should further extension be made
2.7	Liquidated damages for non-completion	24.2	2.3	11.3	
2.8	Repayment of liquidated damages	24.2	—	11.4	
2.9	Practical completion	17.1	2.4	12.1	In ACA, the concept of 'taking over' is similar to 'practical completion'
2.10	Defects liability	17.2	2.5	12.2	

Contracts compared

Table 2.1
IFC Clauses compared with those of other common standard contracts—*cont.*

IFC clause	Description	JCT80 clause	MW80 clause	ACA2 clause	Comment (on IFC84 unless otherwise stated)
3	**Control of the works**				
3.1	Assignment	19.1	3.1	9.1	
3.2	Subcontracting	19.2	3.2	9.2	
3.3.1	Named persons as subcontractors	—	—	9.4 9.5	The IFC and ACA provisions for named subcontractors are not really comparable
3.4	Contractor's person-in-charge	10	3.3	5.2	
3.5	Architect's instructions	4.1	3.5	8.1	There is no provision for confirmation of oral instructions
3.6	Variations	13	3.6	8.1	
3.7	Valuation of variations and provisional sum work	13.4 13.5	3.6 3.7	8.2	
3.8	Instructions to expend provisional sums	13.3	3.7	8.1	
3.9	Levels and setting out	7	—	—	
3.10	Clerk of works	12	—	—	
3.11	Work not forming part of the contract	29	—	10	
3.12	Instructions as to inspection; tests	8.3	—	8.1	

Contracts compared

3.13	Instructions following failure of work, etc	—	—	—	Not exactly comparable with powers in other contracts
3.14	Instruction as to removal of work, etc	8.4	—	8.1	
3.15	Instructions as to postponement	23.2	—	11.8	
4	**Payment**				
4.1	Contract sum	14	—	15	
4.2	Interim payments	30.1	4.2	16.1 16.2	
4.3	Interim payment on practical completion	—	4.3	—	
4.4	Interest in percentage withheld	30.5	4.2	16.4 16.5	MW clause 4.2 is not exactly comparable with similar clauses in other contracts
4.5	Computation of adjusted contract sum	30.6	4.4	19.1	
4.6	Issue of final certificate	30.8	4.4	19.2	
4.7	Effect of final certificate	30.9	—	19.5	
4.8	Effect of certificates other than final	30.10	—	19.5	
4.9	Fluctuations	38, 39 & 40	4.5	18	Under ACA terms, the fluctuations clause is optional. A single index is used and only 80% is payable to the contractor
4.10	Fluctuations; named persons				
4.11	Disturbance of regular progress	26.1	—	7.1	

13

Table 2.1
IFC Clauses compared with those of other common standard contracts—*cont.*

IFC clause	Description	JCT80 clause	MW80 clause	ACA2 clause	Comment (on IFC84 unless otherwise stated)
4.12	Matters referred to in clause 4.11	26.2	—	7.1	
5	**Statutory obligations, etc**				
5.1	Statutory obligations, notices, fees, and charges	6	5.1	1.6 1.7	
5.2	Notice of divergence from statutory requirements	6	5.1	1.6	
5.3	Extent of contractor's liability for non-compliance	6	5.1	1.7	
5.4	Emergency compliance	6	—	—	
5.5	Value added tax	15	5.2	16.7	
5.6	Statutory tax deduction scheme	31	5.3	24	See Inland Revenue publication IR/14/15, Construction Industry Tax Deduction Scheme (1980 edn)
5.7	Fair wages	19A	5.4	—	
6	**Injury, damage and insurance**				
6.1	Injury to persons and property and employer's indemnity	20	6.1 6.2	6.3	

14

Contracts compared

6.2	Insurance against injury to persons and property	21	6.1 6.2	6.3	
6.3A	Insurance in joint names of employer and contractor (new buildings)	22A	6.3A	6.4	
6.3B	Sole risk of employer (new buildings)	22B	—	—	
6.3C	Sole risk of employer (existing structure)	22C	6.3B	6.4	
7	**Determination**				
7.1	Determination by employer	27.1	7.1	20.1	
7.2	Contractor becoming bankrupt, etc	27.2	7.1	20.3	
7.3	Corruption—determination by employer	27.3	—	—	
7.4	Consequences of determination under clauses 7.1-7.3	27.4	7.1	22.1	
7.5	Determination by contractor	28.1	7.2	20.2	
7.6	Employer becoming bankrupt, etc	28.1	7.2	20.3	
7.7	Consequences of determination under clauses 7.5 or 7.6	28.2	7.2	22.2	
7.8	Determination by employer or contractor	28.1	—	21	JCT80 is not exactly comparable, since only the contractor has a right to determine when works are suspended for a period specified in the appendix. IFC84 is closer to ACA in this respect

Contracts compared

Table 2.1
IFC Clauses compared with those of other common standard contracts—*cont.*

IFC clause	Description	JCT80 clause	MW80 clause	ACA2 clause	Comment (on IFC84 unless otherwise stated)
7.9	Consequences of determination under clause 7.8	—	—	22.3	
8	**Interpretation**			23.2	
8.1	References to clauses, etc	1.1	—	—	
8.2	Articles to be read as a whole	1.2	—	—	
8.3	Definitions	1.3	—	—	
8.4	The architect/supervising officer	Art 3A	—	—	
8.5	Priced specification or priced schedules of work	N/A	—	—	

3 Contract documents and insurance

3.1 Contract documents

3.1.1 Types and uses

IFC84 is designed to be a very flexible contract to cover a broad range of work and costs. To give effect to this intention, there is a number of possible combinations which can constitute the 'contract documents'.
What are contract documents? They are those documents which give legal effect to the intentions of the parties. In principle, the contract documents may consist of, and contain, whatever the parties wish. IFC84 sets out the options in the second recital:
— the contract drawings and the specification, priced by the contractor, or
— the contract drawings and the schedule of work, priced by the contractor, or
— the contract drawings and the bills of quantities, priced by the contractor, or
— the contract drawings and the specification and the sum the contractor requires for carrying out the works.
One of these options, together with the agreement and conditions annexed to the recitals, forms the contract documents. They must all be signed by, or on behalf of, the parties.
You should study the options carefully in order to arrive at the most suitable combination for a particular project.

The contract drawings and the specification priced by the contractor

this combination is usually appropriate for relatively small works or for works of a simple nature. It must be remembered that, if you do not provide bills of quantities, the contractor will have to take off his own quantities in order to arrive at a tender sum. The prices eventually put to the specification will inevitably be somewhat rough and ready unless your specification is a model of clarity. Since the priced specification will be the basis of the valuation of variations (clause 3.7), there could be pitfalls for contractor and employer alike.

The contract drawings and the schedule of work priced by the contractor

Useful for work which is somewhat more complicated than the last example but does not warrant full bills of quantities. Again, the contractor will be obliged to take off his own quantities, and the employer, or the building industry in general, will in effect bear the cost. A schedule of work demands that there is agreement on the order in which the work will be carried out. Even on a relatively simple job, the contractor may be able to devise cheaper and more efficient methods of arriving at the result than those you envisage. Some form of two-stage tendering procedure is indicated to allow such a contribution by the contractor before the schedule is completed. Whether that is warranted will depend on all the circumstances. A priced schedule is more comprehensible, for the purpose of valuations, than a priced specification, but it requires the utmost clarity of thought to prepare properly.

The contract drawings and the bills of quantities priced by the contractor

If the size of the job justifies it, this must be the most satisfactory all-round combination. It is tried and tested, and each party knows not only the price for the whole job, but also the cost of variations. Monthly valuations are simplified, and there is less likelihood that the contractor will perpetrate some terrible undetected mistake at tender stage.

The contract drawings and the specification and the sum the contractor requires for carrying out the works

This combination requires the contractor to take off his own quantities and supply a total price, which will become the contract sum. He is not required to price the specification but, instead, to supply the employer with a contract sum analysis of the stated sum or a schedule of rates on which the stated sum is based. The first point to note is that the contract does not include either the contract–sum analysis or the schedule of rates as a contract document. This appears to be a serious miscalculation, although clearly it is intentional (second recital,

3.1 Contract documents

alternative B). There seems to be no good reason why either document should not be as important as, say, the priced specification in the first option. It is suggested that, if you intend to advise the employer to use alternative B, you should also amend the second recital to include the pricing documents as contract documents. The contract–sum analysis is stated (clause 8.3) to mean an analysis of the contract sum in accordance with the stated requirements of the employer. The definition purposely leaves room for you to require the contractor to provide the analysis in any form you require, presumably including bills of quantities. The schedule of rates is more familiar and may be thought more useful than the schedules of work in another option. It is likely that this option will be used in most cases where bills of quantities are not prepared.

The contract drawings are to be noted by number in the designated place in the first recital. There tends to be some disagreement over the number and type of drawings to be designated contract drawings. They must be sufficiently detailed to show the location and extent of the work; those from which the contractor obtained information to submit his tender; and related to the other contract documents.

Ideally, the contract drawings should include every drawing prepared for the work. In practice, this is not always possible, but if drawings and specification are being used to obtain tenders, the drawings must be detailed enough to allow the contractor to carry out his own taking off. If bills of quantities are provided, small details may be omitted.

The further drawings and details which you are to provide under clause 1.7 are not contract documents. If they show different or additional or less work and materials than are shown on the contract documents, the contractor will be entitled to a variation.

All the contract documents must be signed and dated by both parties. That means every separate drawing or separate piece of paper, but not every sheet of the specification or bills of quantities—signing the cover or the last page is sufficient. An endorsement on each document should read: 'This is one of the contract documents referred to in the agreement dated . . .' or other words to the same effect.

3.1.2 Importance and priority

The significance of the contract documents has already been discussed briefly. If a dispute arises and it is necessary to discover what was agreed between the parties, the arbitrator or the court will look at the contract documents.

In a number of places throughout the contract, reference is made to work being or not being in accordance with the contract (eg clause 3.14) That means in accordance with what is contained in the contract

Contract documents and insurance

documents. A problem arises if the documents are in conflict. Clause 1.2 sets out rules which are to be followed depending on which combination of documents you and the employer have chosen.

In order to decide on the quality and quantity of work agreed to be carried out for the contract sum, you must look at the particular documents.

Drawings and specification (or schedules of work)
They must be read together, provided that no quantities are shown. If there is a conflict, the drawings are to be given preference over the specification or schedules of work. If quantities are shown for some items, those quantities will prevail.

Drawings and bills of quantities.
The quality and quantity of work shown in the bills of quantities will prevail.

So if the drawings show a total of 60 holding-down straps and the specification states that 50 holding-down straps are required, the contractor may quite correctly price for 50 holding-down straps. If, at a later stage in the contract works, you decide that you do need 60 straps as shown on the drawings, the employer will have to pay for the additional 10.

If, on the other hand, the straps are mentioned in the specification without being quantified, the contractor will be deemed to have priced for 60 straps. This clause will be welcomed by contractors, because it clarifies a situation which can be a source of dispute when the architect for a particular project maintains that the contractor should price for everything, whether it is mentioned in the specification or on the drawing. This position, however, still applies in this contract unless quantities are mentioned. In practice, the result is likely to be that you will have to take great care when you prepare the specification, or the employer will face large bills for additional work.

Clause 1.3 provides that nothing in the bills of quantities/the specification/the schedules of work (as appropriate) will override or modify the application or interpretation of anything in the articles, conditions, supplemental conditions, or appendix. In effect, it means that you cannot amend anything in the printed form by inserting a clause in the bills of quantities, etc. This is so even if the insertion is written in ink and signed by both parties. The way to amend the printed form is to do so on the form itself and have the amendment signed or initialled by the parties. Alternatively, suitably amended or special clauses can be annexed to the form, duly signed by the parties, or clause 1.3 itself can be deleted. Were it not for this clause, the general law would give effect to any amendments which were contained in the bills, etc. This clause is therefore of great significance.

3.1 Contracts documents

3.1.3 Errors

Clause 1.4 further stresses the importance of having a thoroughly prepared set of contract documents and of using the same care to include anything which you may issue to the contractor to assist him in carrying out the works.

You are obliged to issue instructions (see section 4.1.4) to correct any of the following:

— inconsistencies which occur within any one of the contract documents or between the contract documents: this obligation extends to inconsistencies which may occur following your issue of further information under clause 1.7 or levels and setting-out information under clause 3.9

— errors in description or quantity or omissions of any item in any of the contract documents

— errors or omissions in the particulars which the employer provides in relation to a person who is named in accordance with clause 3.3.1 (see section 8.2.2)

— departures from the method of preparation of the bills of quantities referred to in clause 1.5: this states that the bills of quantities must be prepared in accordance with the Standard Method of Measurement of Building Works, sixth edition, except where certain items are expressly stated to have been measured in a different way. This is always a fruitful source of claims by the contractor.

None of these errors, inconsistencies, or departures will invalidate the contract, but if an instruction changes the quality or quantity of work as understood from clause 1.2 or changes any clause 3.6.2 obligations or restrictions, a variation will result.

There is a crumb of comfort to be gained from the fact that it is established law that, if something is omitted from the bills or specification which it is quite clear to everybody should be there and is necessary, the contractor will be deemed to have included it in his price. There is scope for dispute in applying this principle. The contractor will always maintain that it certainly was not clear to him, or that he assumed that the employer was going to employ someone else to do or supply the thing in question. In the context of a particular contract, however, it should be possible to make a fair decision on any particular item.

An example will clarify the point. Assume that the bills of quantities provide for the contractor to supply damp–proof course of a particular quality and in a given quantity, but the item requiring him to lay it has been inadvertently omitted. He will be deemed to have included the laying in his price because it is clear to everyone that laying is required. On the other hand, if you fail to include the supply and fixing of, say,

Contract documents and insurance

door furniture, the contractor may have a point in claiming that he assumed that the employer wanted to carry out this operation himself. Moreover, he will have had no information at all on which to base any price, and he would be entitled to his extra costs. In practice, the extreme situation should never arise, because the contractor would certainly query the omission of door furniture at tender stage.

3.1.4 Custody and copies

Clause 1.6 makes it quite clear that the contract documents must remain in the custody of the employer. There is a proviso that the contractor must be allowed to inspect them at all reasonable times, but this provision appears to be redundant in the light of the fact that you must provide him with a copy of the documents certified on behalf of the employer. The employer will expect you to check that the copy is the same as the original. Very often, two sets of documents are prepared and all are signed by both parties, but it is not strictly necessary. You would be wise to let the employer check that the copy is the same as the original and sign the certificate himself, since that is the employer's responsibility. The certificate is usually inscribed on each document, including the drawings. Some architects favour a very elaborate pseudo-legal turn of phrase, but it is sufficient to state, 'I certify that this is a true copy of the contract document'.

The clause imposes a further duty on you to supply the contractor with two copies (uncertified) of each of the contract documents.

The further drawings and details which you are to supply under clause 1.7 are not contract documents. They are intended merely to amplify the information contained in the contract documents. Your obligation is to supply only such drawings and details as are reasonably necessary to enable the contractor to carry out and complete the works in accordance with the conditions: ie among other things, to complete by the stipulated completion date. The contractor has no claim for extension of time or loss and/or expense if you fail, unless he has made a specific written application at the proper time in accordance with clause 2.4.7 or clause 4.12.1 (see section 10.2.4).

3.1.5 Limits to use

Clause 1.8 contains safeguards for both contractor and architect. It prohibits the use for any purpose other than the contract of any documents (contract documents or others) issued in connection with the contract. It also prohibits the employer, the architect, and the quantity surveyor from using any of the contractor's rates or prices in the contract documents, the contract sum analysis, or the schedule of

3.1 Contract documents

rates for any purpose other than the contract. None of the contractor's rates or prices must be divulged to third parties.

The prohibition against the use of documents by the contractor simply states expressly what is understood from the general law with regard to your copyright.

The prohibition against divulging the contractor's rates is designed to protect the contractor's most precious possession—his ability to tender competitively and thus secure work. To divulge his rates to a competitor is probably one of the most harmful things you could do to any contractor. In practice, it is extremely difficult for the contractor to ensure that his prices are not used, for example, by the quantity surveyor to help him estimate some other current job.

There is no requirement for the contractor to return any drawings or details at the end of the contract. There is nothing to prevent you from asking for them in order to be sure that they will not be used for any other purpose, but you cannot require the contractor to return his certified copy of the contract documents, which he will need for his own records.

3.2 Insurance

3.2.1 Indemnity

Under clause 6.1, the contractor assumes liability for, and indemnifies the employer against, any liability arising out of the carrying out of the works in respect of the following:

—   personal injury or death of any person, unless due to act or neglect of the employer or of any person for whom he is responsible: the persons for whom the employer is responsible will include anyone employed and paid by him, such as directly employed contractors, the clerk of works, and yourself.

—   injury or damage to any kind of property, provided that it is due to the negligence, omission, or default of the contractor or of any subcontractor or any of their respective employees or agents: the contractor's liability is limited, as compared to his liability for personal injury or death, since he must be at fault for the indemnity to be operative. He has no liability under this clause for any loss or damage which is at the sole risk of the employer under clauses 6.3B or 6.3C.

From the point of view of the employer, he is responsible for the injury or death of any person only if the injury or death was caused by his or his agents' act or neglect. He is responsible for all loss or damage to property unless caused by the contractor's or subcontractor's default.

In practice, if any claim does arise, it is likely to be the employer who will be sued. In turn, he will join the contractor as a third party in any action and claim an indemnity from him under this clause.

3.2.2 Injury to persons and property

Clause 6.2.1 requires the contractor to take out insurance (unless his existing insurances are adequate) to cover his liabilities under clause 6.1. This requirement is stated to be without prejudice to his liability to indemnify the employer under that clause. Thus the fact that the contractor has taken out or maintains the appropriate insurance cover does not affect his liabilities. If, for some reason, the insurance company refused to pay in the case of an incident, the contractor would be obliged to find the money himself. The contractor must ensure that all subcontractors maintain similar insurance in respect of their own and the contractor's liability.

The insurance cover must be for a sum not less than whatever is stated in the appendix to the contract for any one occurrence or series of occurrences arising out of one event. The insurance against claims for personal injury or death of an employee or apprentice of the contractor or a subcontractor must comply with the Employer's Liability (Compulsory Insurance) Act 1969.

You have the right (clause 6.2.2) to inspect documentary evidence that the contractor and his subcontractors are maintaining proper insurance cover, and in particular to inspect policies and premium receipts. You are not to exercise this right unreasonably or vexatiously: you are likely to enforce it at the beginning of the contract and at the time of any required premium renewal. It is always wise to retain the service of an insurance broker in connection with all the insurance provisions of the contract. It is for the employer to appoint him on your advice. He should inspect the relevant documents and confirm to you in writing that they comply with the contract requirements.

If the contractor fails to insure under clause 6.2.1, the employer has the right (clause 6.2.3) to take out the appropriate insurance himself and deduct the amount of any premium from monies due or to become due to the contractor. Alternatively, the employer may recover the amount from the contractor as a debt. It is essential that the employer exercises that right, and such is the importance of maintaining continuous insurance cover, you are probably justified in taking action on the employer's behalf immediately you discover that the contractor has defaulted. The cover should be effected through and on the advice of the employer's broker. You must immediately confirm your actions to the contractor and the employer. The exact circumstances may vary widely, but the letters in Figs 3.1 and 3.2 are examples.

3.2 Insurance

Fig 3.1
Architect to contractor if contractor fails to maintain clause 6.2.1 insurance cover

```
Dear Sir

I refer to my telephone conversation with your
Mr [insert name] this morning and confirm that you
are unable to produce the insurance policy, premium
receipts, or documentary evidence that the
insurances required by clause 6.2.1 are being
maintained.

In view of the importance of the insurance and
without prejudice to your liabilities under clause
6.1 of the conditions of contract, [the employer]
is arranging to exercise his rights under clause
6.2.3 immediately.  Any sum or sums payable by him
in respect of premiums will be deducted from any
monies due or to become due to you or will be
recovered from you as a debt.


Yours faithfully



Copy: Employer
      Quantity surveyor
```

Fig 3.2
Architect to employer if contractor fails to maintain clause 6.2.1 insurance cover

Dear Sir

The contractor is unable to provide evidence to show that he is maintaining the insurances required by clause 6.2.1 of the conditions of contract.

In view of the importance of the insurance, I have taken action on your behalf, under clause 6.2.3, and instructed your broker to provide the necessary cover effective from today. You are entitled to deduct the amount of the premium from your next payment to the contractor or, alternatively, you may wish to recover it as a debt. [*Add if appropriate*] In this instance, simple deduction would appear to be the easiest method of recovery.

A copy of my letter to the contractor, dated [*insert date*] is enclosed for your information.

Yours faithfully

3.2 Insurance

3.2.3 Things which are the liability of the employer

Clause 6.2.4 provides for insurance against damage caused by the carrying out of the works when there is no negligence or default by any party. The clause is operative only if a provisional sum has been included in the contract documents for the purpose. The contractor must maintain the insurance in the joint names of the employer and the contractor. The amount of cover must be specified in the contract documents. Liability envisaged by this clause covers damage to any property, other than the works, caused by collapse, subsidence, vibration, weakening or removal of support, or lowering of ground water. Not covered is damage which is caused by the contractor's or subcontractor's negligence, omission, or default or that of their respective employees or agents; due to errors or omissions in the architect's design; reasonably foreseen to be inevitable; at the sole risk of the employer under clause 6.3B or clause 6.3C; or due to nuclear risk, war risk, or sonic booms.

The contractor must obtain your approval of the insurers he proposes to use, but not, apparently, to the amount of the premium. One method of sorting this out is for you to place the matter in the hands of the employer's broker. You will need his advice for the purpose of arriving at a provisional sum before the contract is let. If you later indicate to the contractor that you would be prepared to approve the insurer recommended by the broker, the contractor will probably be delighted to place the insurance with whomsoever you wish. The device of the 'double letter' (Figs 3.3. and 3.4) is useful in such circumstances, as it leaves the responsibility where it belongs—with the contractor. The two letters are sent at the same time.

The contractor must deposit the policy and premium receipts with you. The amounts paid by the contractor in premiums under this clause are to be added to the contract sum. If the contractor fails to insure or to maintain the insurance under this clause, the employer may take out the insurance himself, in which case nothing is added to the contract sum under this clause. Since the employer, or you on the employer's behalf, know what insurance is needed and know which insurer you wish to provide the cover, there may be something to be said for amending this clause to remove the provisional sum and make the employer responsible for insuring under this clause.

There is a general proviso, contained in clause 6.2.5, stating that the contractor shall not be liable to indemnify the employer or to insure against damage to the works, the site, or any property due to nuclear perils and the like. This clause overrides anything contained in clauses 6.1 and 6.2.

Fig 3.3.
Architect to contractor regarding clause 6.2.4 insurance (double letter 1)

```
Dear Sir

I should be pleased if you would inform me of the
name of the insurers with whom you intend to place
insurance in accordance with clause 6.2.4 of the
conditions of contract.  The contract provides
that I must give my approval before you proceed
to place the insurance.


Yours faithfully
```

3.2 Insurance

Fig 3.4
Architect to contractor regarding clause 6.2.4 insurance (double letter 2)

```
WITHOUT PREJUDICE


Dear Sir

With regard to the insurer whose name you are
required to submit to me under clause 6.2.4 of
the conditions of contract, if you were to suggest
```
[*insert name of insurers recommended by the employer's broker*] ```I should be prepared to approve
them.

Yours faithfully
```
```

Contract documents and insurance

3.2.4 Clause 6.3 perils

Clause 6.3 deals with the insurance of the works against clause 6.3 perils. Clause 6.3 perils are
Fire, lightning, explosion, storm, tempest, flood, bursting or overflowing of water tanks, apparatus, or pipes, earthquake, aircraft and other aerial devices or articles dropped therefrom, riot and civil commotion, excluding any loss or damage caused by ionising radiations or contamination by radioactivity from any nuclear fuel, radioactive toxic explosive or other hazardous properties of any explosive nuclear assembly or nuclear component thereof, pressure waves caused by aircraft or other aerial devices travelling at sonic or supersonic speeds.
The clause is divided into three parts, two of which are to be deleted. Each part is applicable to a particular situation as follows: 6.3A, a new building where the contractor is required to insure; 6.3B, a new building at the sole risk of the employer; 6.3C, alterations or extensions to an existing building.

3.2.5 A new building where the contractor is required to insure

The provision is relatively straightforward and provides for the contractor to insure in the joint names of the employer and himself against loss or damage by clause 6.3 perils. The insurance must be for the full reinstatement value, plus whatever percentage is given in the appendix to cover professional fees.
To be included are all executed work, and all unfixed materials and goods, delivered to, placed on, or adjacent to the works and intended for incorporation. Temporary buildings, plant, tools and equipment owned or hired by the contractor or any subcontractor are excluded. The insurance cover must be kept in force until the date of issue of the certificate of practical completion, however long that may be.
The contractor must obtain your approval or the approval of the employer to the insurers before he proceeds with taking out the policy. If he fails to insure or to maintain the insurance, the employer may take out insurance himself to cover the clause 6.3 perils and may deduct the amount of the premiums from monies due or to become due, or he may recover it from the contractor as a debt. In any case, the policy and the premium receipts are to be held by the employer.
If the contractor maintains his own insurance policy which provides the same degree and type of insurance as required by this clause, it will serve to discharge the contractor's obligations under clause 6.3A.1. Such a policy must be endorsed to show the employer's interest. Documentary evidence of the endorsement, the policy, and premium

receipts must be produced for inspection whenever you or the employer reasonably and not vexatiously require it. If the contractor defaults in maintaining his own insurance, the employer's powers under clause 6.3A.2 to take out insurance can be invoked (clause 6.3A.3).

Clause 6.3A.4 provides for the procedure if damage is caused by clause 6.3 perils. The contractor need do nothing until the insurers have accepted the claim. On acceptance, he must restore damaged work, replace or repair unfixed materials, remove and dispose of debris, and then proceed with the works as before. He is entitled to be paid all the money received from insurances (except for the percentage to cover professional fees) on certificates in the normal way. He is entitled to no other money for restoration and repair work. So if there is any under-insurance, the contractor has to make good the shortfall at his own expense.

3.2.6 A new building at the sole risk of the employer

Under clause 6.3B.1, the matters for which the contractor is required to take out insurance under clause 6.3A.1 are at the sole risk of the employer. If the employer is a local authority, there is no requirement to insure. Otherwise, to protect the contractor's interests, the employer must maintain a policy covering all the matters which are at his risk except professional fees. The contractor is entitled to inspect the policy and premium receipts at any time by simple request. If the employer fails to produce a receipt, the contractor may insure against clause 6.3 perils in the employer's name and is entitled to have the amount of any premium he pays added to the contract sum on production of a receipt.

Clause 6.3B.3 deals with the procedure to be adopted if damage occurs from clause 6.3 perils. On discovery of the damage, the contractor must immediately give notice in writing to both the employer and yourself, stating its nature, extent, and location. He must then carry out restoration, repair, and removal of debris and proceed with the works in the normal way. All work caused by the damage must be treated as though it was a variation required by your instruction and the contractor be paid accordingly. You must take no account of the damage when you compute any amount normally payable to the contractor under the contract.

In contrast to clause 6.3A, this clause tries to ensure that the contractor suffers no loss of any sort if the work is damaged by clause 6.3 perils.

3.2.7 Alterations or extensions to an existing building

In the case of an existing building, there is no option for the contractor to insure. Even if the employer is a local authority, he must insure the existing building, the works, and all unfixed materials, etc, as before, unless the clause is deleted. The contractor's temporary buildings, etc, are at his own risk. Professional fees are not mentioned, so although they are the employer's responsibility, he is not required to insure for their value. If he is wise, of course, he will insure them, and it is worth while making the point to the employer.

If the employer fails to produce receipts to the contractor on request, the contractor has quite extraordinary powers. Not only may he insure the existing building, works, and unfixed materials in the employer's name, he also has a right of entry sufficient for a survey and inventory of the existing property in order to obtain the required insurance cover. On production of premium receipts, the contractor is entitled to have the amounts shown added to the contract sum. If it has been necessary to carry out a survey or make an inventory, most contractors will expect to be reimbursed any additional costs involved. Since there is no contractual provision, it will be a matter for the employer.

Under clause 6.3C.2, the contractor is required to give written notice of any discovered damage, to both you and the employer. If neither party then determines the employment of the contractor (see section 12.1.6), the contractor must proceed to restore and repair and continue with the works. Strangely, the removal of debris is made subject to your instructions on the matter. This can only be because some of the debris may be from existing structures, about which special instructions are appropriate, although the same argument could be applied to the reinstatement of existing structures, for which your particular instructions are not required.

All work caused by the damage is to be treated as though it was a variation required by your instruction and the contractor paid accordingly. No account of the damage must be taken when you compute amounts normally payable to the contractor under the contract.

3.3 Summary

Contract documents

— contract documents are the only evidence of the contract
— they may consist of whatever the parties agree
— drawings prevail over the specification
— any quantities prevail over the drawings
— printed conditions prevail over all
— errors, inconsistencies, and departures are to be corrected by architect's instruction and a variation allowed if appropriate
— items missing from the specification may be deemed to be included if it is obvious to all that they should be there

3.3 Summary

— contract documents must be kept by the employer with a true copy to the contractor
— no other drawing or detail is a contract document
— contract documents and any other documents issued for the contract must not be used for any other purpose
— contractor's rates must not be divulged or used for another purpose

Insurance

— employer's indemnity covers personal injury and death and damage to property, subject to certain exceptions
— contractor must insure to cover the indemnities
— employer has the right to insure if the contractor fails
— special insurance can be taken out to cover instances where there is no default by any party
— either party may insure new work against clause 6.3 perils
— only the employer may insure existing work against clause 6.3 perils.

4 The architect's authority and duties

4.1 Authority

4.1.1 General

The extent of your authority depends on your agreement with the employer. If you are wise, you will have entered into a formal written contract, preferably incorporating the terms of the RIBA *Architect's appointment*. Your powers and duties under IFC84 (see Table 4.1) flow directly from your agreement with the employer, not from the building contract itself, to which you are not a party. It follows that if you fail properly to carry out your duties, the employer, but not the contractor, can take legal action against you under the terms of your contract with him. It is possible for the contractor to bring an action against you in tort for negligence but he must show that you owed a legal duty of care to him, and that you were in breach of that duty, and that, by reason of the breach, he suffered loss or damage.

Generally, contractors find it easier to take action against the employer under the contract, the employer in turn taking action against you. If you fail to carry out your duties under the contract, this is generally a default for which the employer will be held responsible.

Traditionally, your role changes when the contract between employer and contractor is signed. Up to that time, you have been acting as agent, in a limited capacity, for the employer. After the signing, you assume a dual and difficult role. You are still an agent of the employer, but you are also charged with seeing that the terms of the contract are administered fairly—'without fear or favour' is the usual term.

4.1 Authority

'Without favour' is absolutely right, but 'without fear' is obsolete nowadays. It used to be thought that, when you carried out your duties under the contract, you were immune from any action against you in negligence by either party. If such a state ever really existed, it was changed in 1974 by the celebrated case of *Sutcliffe* v *Thackrah*. There is now no doubt that you are open to an action for negligence on every decision you take. You are not acting in a 'quasi-arbitral' capacity. Notwithstanding that, you are still required to act fairly between the parties in your administration of the contract. The situation is totally unrealistic, because there may be instances when you have to decide whether you yourself are in default and to act accordingly—for example, in cases of claims for extension of time or for loss and/or expense. If you issue instructions late, you are duty bound, if the contractor makes a proper claim under the contract, to award an extension or ascertain a financial claim as the case may be. The employer may well be able to recover that loss from you.

It is important to note that, as far as the contractor is concerned your authority is stated in the contract, and is neither more nor less. Thus if, for example, you attempt to issue an instruction not empowered by the contract, the contractor need not carry it out. Indeed, if he does carry out an instruction which is not empowered by the contract, the employer probably has no liability (but see section 4.1.3). If your instruction is empowered, the contractor need not worry whether you have the employer's consent: he may carry it out and the employer is bound.

Two simple examples should make the position clear. Assume that your contract with the employer is on the terms set out in the *Architect's appointment*. Part 3.3 provides that you are not to make any material alteration, addition, or omission from the approved design without the employer's consent unless it becomes necessary during construction for constructional reasons and you inform the employer without delay—in other words, if an emergency arises. If, as a first example, you issue an instruction to vary the quality of the electrical fittings without the employer's consent, you will be in breach of your contract with him. The contractor can carry out the work, because you are empowered to issue such instructions by IFC84, and the employer must pay if any extra cost is involved. The employer can recover any costs from you. As a second example, suppose you inform the contractor that he may take possession of the site two weeks before the appointed date. The contract gives you no authority to vary its terms—indeed, only the parties to the contract can do that. The contractor would take possession at his peril, the employer would have no liability and, if it came to that, you could face action from the contractor in tort.

If you are on the staff of the employer (for example, in a local authority), the position is much less clear. The contractor may quite rightly

Table 4.1
Architect's powers and duties under IFC84

Clause	Power/duty	Precondition/comment
1.4	**Duty** Issue instructions as to inconsistencies, errors, or omissions in or between the contract documents, drawings, etc	The power to correct errors in description or quantity applies only to items and not to prices
1.4	**Duty** Value the correction under clause 3.7 if instruction changes quality or quantity of work	
1.6	**Duty** Provide contractor with one copy of contract documents certified on behalf of employer and two further copies of contract documents and specification/schedule of work/bills	
1.7	**Duty** Provide contractor with two copies of further drawings or details necessary for the proper carrying out of the works	
1.8	**Duty** Not to divulge to third parties or use any of the contract rates for prices except for purposes of the contract	
1.10	**Power** Consent in writing to removal of unfixed materials or goods delivered to, or placed on or adjacent to, the works	Contractor's request must be for good reason; consent must not be unreasonably withheld
2.3	**Duty** Make in writing a fair and reasonable extension of time for completion as soon as he is able to estimate length of delay	It must become apparent that the progress of the works is being, or is likely to be, delayed *and* Contractor must forthwith give written notice to the architect of the cause of the delay *and* Architect must form the opinion that completion is likely to be, or has been, delayed beyond the original or extended completion date *and* Reasons for delay must be a relevant event as listed in clause 2.4

4.1 Authority

2.3	**Power** Make in writing a fair and reasonable extension of time for completion even if contractor has not given notice	He can do this at any time up to 12 weeks after date of practical completion
2.3	**Power** Review extensions of time granted	This is implied, and the power should always be exercised in order to preserve the employer's right to liquidated damages if there has been some default for which the employer is responsible
2.6	**Duty** Issue a certificate of non-completion	If contractor fails to complete the works by the date for completion or within extended period
	Duty Cancel clause 2.6 certificate in writing and issue further certificate as necessary	If an extension of time is granted subsequent to the issue of the former certificate
2.9	**Duty** Certify the date when practical completion is achieved	When he is of the opinion that this is so
2.10	**Power** Issue instructions that defects, etc, be not made good	If the employer consents. An appropriate deduction must be made from the contract sum
	Duty Certify discharge of contractor's defects liability	When he is of the opinion that the contractor has discharged his obligations
3.2	**Power** Consent to employment of subcontractor	Application from contractor Consent must not be unreasonably withheld
3.3.1	**Duty** Issue instructions changing particulars or omitting the work or substituting a provisional sum	Where contractor is unable to subcontract in accord with particulars and so informs architect
3.3.1	**Power** Issue instruction requiring named subcontractor work to be carried out by another person	Must be done before contractor notifies architect of his entering into named subcontract
3.3.2	**Power** Instruct that provisional sum work be carried out by named person	In an instruction as to the expenditure of a provisional sum, the instruction must incorporate a description of the work and relevant tender particulars from NAM/T sections I and II Contractor has right of reasonable objection to named person

37

Table 4.1
Architect's powers and duties under IFC84—cont.

Clause	Power/duty	Precondition/comment
3.3.3	**Duty** Issue necessary instructions should employment of named person be determined before completion of subcontract work	Instruction must: *either* name another person and give description of work and relevant particulars; *or* instruct contractor to make his own arrangements; *or* omit the work
3.5.1	**Power** Issue written instructions **Power** Require compliance with an instruction by written notice to contractor	If the contractor fails to comply, the employer may employ others
3.5.2	**Duty** Specify in writing the contract clause empowering issue of an instruction	On contractor's written request
3.6	**Power** Issue variation instructions and sanction in writing any unauthorised variation made by the contractor	The variation ordered must fall within the definition in 3.6.1 or 3.6.2. Variations must be valued by the QS if no prior agreement between employer and contractor
3.8	**Duty** Issue instructions as to expenditure of provisional sums	
3.9	**Duty** Determine levels required for execution of the works and provide contractor with information necessary to enable him to set the works out **Power** Instruct contractor not to amend setting-out errors	The information must be conveyed by means of accurately dimensioned drawings If the employer consents. An appropriate deduction must be made from contract sum
3.12	**Power** Issue written instruction requiring opening up of work, etc, for inspection or testing	Cost to be added to contract sum unless results show the work, etc, is not in accordance with the contract

4.1 Authority

3.13.1	**Duty** Issue instructions following failure of work, materials, or goods	Where architect has *either* not received contractor's proposals within seven days of discovery of failure, *or* is not satisfied with contractor's proposed actions, *or* safety considerations or statutory obligations require urgent action
3.13.2	**Power** Withdraw or modify instruction issued under clause 3.13.1	If contractor objects to compliance within 10 days of receipt of 3.13.1 instructions and architect accepts contractor's reasons
3.14	**Power** Issue written instructions requiring removal of work, materials, or goods not in accordance with the contract	
3.15	**Power** Issue instructions as to postponement of any *work*	
4.2	**Duty** Certify interim payments at monthly intervals calculated from date of possession stated in appendix	This is subject to any agreement between employer and contractor as to stage payments. Different intervals may be specified in the appendix. Interim valuations are to be made by the QS whenever architect considers them to be necessary to ascertain amount to be certified The amounts to be included are 95 per cent of the total value of: work properly executed, including variations and fluctuations adjustment if appropriate, and materials and goods for incorporation which have been reasonably and properly and not prematurely delivered to site (they must be adequately protected against the weather, etc); and 100% of ascertained disturbance claims, fees, insurance payments etc.
4.2.1	**Power** Include the value of off-site goods and materials in interim certificates	
4.3	**Duty** Certify interim payment to the contractor of 97½% of total amount to be paid to contractor	Within 14 days after the certified date of practical completion
4.5	**Power** Instruct contractor to send to QS documentation needed for computing adjusted contract sum. **Duty** Send copy of computations of adjusted contract sum within period of final measurement and valuation stated in appendix	

The architect's authority and duties

Table 4.1
Architect's powers and duties under IFC84—*cont.*

Clause	Power/duty	Precondition/comment
4.6	**Duty** Issue final certificate	This must be done within 28 days of sending the computations of the adjusted contract sum to the contractor *or* of the date of issue of the clause 2.10 defects liability certificate, whichever is the later
4.11	**Duty** Ascertain, or instruct QS to ascertain, the amount of direct loss and/or expense incurred or likely to be incurred by the contractor due to the employer's deferment of site possession under clause 2.2 (if applicable) or to regular progress being materially affected by one or more of the matters specified in clause 4.12	If the contractor makes written application within a reasonable time of its becoming apparent *and* the architect is of opinion that the contractor has incurred or is likely to incur direct loss and/or expense not reimburseable by any other payment under the contract *and* supporting information is supplied
	Power Require from contractor such information to support claim as is reasonably necessary to ascertain amount of direct loss and/or expense	
6.2.2	**Power** Require contractor to produce documentary evidence of his and his subcontractors' insurance under clause 6.2.1	This extends to inspecting the original policy (policies) and premium receipts. The power must be exercised reasonably
6.2.4	**Duty** Approve proposed insurers and accept deposit of contractor's (policies) and premium receipts	A provisional sum must be included in the contract documents for this insurance
6.3A.2	**Power** Approve insurers and accept deposit of policy (policies) and premium receipts	Where the employer does not approve. Subject to clause 6.3A.3 below
6.3A.3	**Power** Require contractor to produce documentary evidence that contractor is maintaining an appropriate policy, and require production of policy itself and premium receipts	Where cover is through general policy maintained by contractor The power must not be exercised unreasonably
6.3C.3	**Power** Issue instructions requiring the contractor to remove and dispose of any debris	Reinstatement and making good of loss or damage and removal and disposal of debris are to be valued under clause 3.6
7.4	**Power** Instruct contractor to remove from the works any temporary buildings, plant, tools, etc	When contractor's employment has been validly determined by the employer

4.1 Authority

assume that you are acting as agent for the employer. In this respect, the contractor need have no regard to whatever standing orders may say, unless they have been specifically drawn to his attention, but may rely on your apparent authority.

4.1.2 Express provisions

Article 3 of IFC84 provides for the insertion of the name of the architect. The person whose name is entered is then the person to whom the contract refers whenever the word 'architect' appears in the conditions. Ideally, the person who is actually to administer the contract should have his name entered. Problems can arise, however, because he may leave the practice, die, or retire. It is generally accepted that the name to be entered will be the name of the practice (ie XYZ & Partners) or of the chief architect in a local authority (eg C. Wren). Most employers and contractors accept this convention and the fact that it will be one of the partners or employees of the practice, or one of the staff of the local authority architect, who will administer the contract. If you are in private practice or local government, you would be wise to ensure that all interested parties, including the employer and the contractor, are informed, at the beginning of the contract, who are the authorised representatives (letter, Fig 4.1). You should also inform them whenever there is a change. The situation is particularly delicate when a client has commissioned you, as one of the partners in a practice, to carry out his project. He will expect that you will oversee every detail. The letter will then serve the purpose of assuring him that, although the practice name is on the contract, you are personally looking after the job. All letters, instructions, certificates, notices, and letters must be signed by the registered architect duly authorised 'for and on behalf of . . .'. It is not usually sufficient to sign your name only, even though you are using headed stationery. The letter may be deemed to be written on your own behalf, which is fine if you are a partner but not so good if you are merely in salaried employment. The unfortunate habit of using a rubber stamp or signing someone else's name and adding your initials should be avoided.

If the architect named in article 3 dies, or if his appointment is terminated, the employer has 14 days in which to nominate a successor. It is now clear that the employer must name a successor. Except where the architect is an official of a local authority, the contractor has the right to object to the nomination. This provision is inserted because the new architect may be someone with whom the contractor has had unsatisfactory dealings in the past. If the employer thinks the contractor's reasons are insufficient, the matter can be decided by arbitration. Clearly, such a situation should be avoided if at all

The architect's authority and duties

Fig 4.1
Architect to contractor, naming authorised representatives

Dear Sir

This is to inform you formally that the architect's authorised representatives for all the purposes of the contract are:

[*insert name*] : Partner in charge of the contract
[*insert name*] : Project architect

The above are the only people authorised to act in connection with this contract until further notice.

Yours faithfully

for and on behalf of [*insert name in contract, which should be the firm's name*]

Copy: Employer
       Quantity surveyor
       Consultants
       Clerk of works

4.1 Authority

possible, not least because of the delaying effect on the contract.

An important provision in this article states that no succeeding architect may disregard or overrule any certificate or instruction given by the previous architect. In the absence of such an express provision, no doubt a similar provision would be implied, because it is essential that the contractor's interests be safeguarded in circumstances which are solely under the control of the employer. If you are the successor architect and you disagree with previous decisions, you would be wise to inform the employer, in writing, of your position, but you cannot alter them.

Certain actions under the contract are left to your discretion—for example, whether to include the value of off-site goods or materials in your certificates (clause 4.2.1 (c)). Whether you include off-site goods or not, you will be deemed to have exercised your discretion. In exercising your discretion, you do not have to account for your actions to the contractor, but you may have to account for them to the employer if things go wrong. So err on the side of caution.

Other clauses call for the contractor to obtain your consent to certain actions—for example, the removal of unfixed materials delivered to, placed on or adjacent to the works, and intended for use therefor (clause 1.10). It is stipulated that your consent must not be unreasonably withheld. If the contractor is not satisfied with your decision, he can refer the matter to arbitration. It is thought that you have an obligation to state your reasons for withholding your consent. If you do not, the contractor may well refer the matter to arbitration anyway. In stating your reasons, be brief.

Matters reserved for your opinion—for example, whether the completion of the works is likely to be delayed (clause 2.3)—are for your opinion alone. You cannot shift the responsibility on to the quantity surveyor, nor must you accept any interference by the employer. Your opinion must not be a whim, however, and it is always prudent to make brief notes for your own files in case your opinion comes up for review during arbitration.

4.1.3 The issue of instructions: general

All instructions which you issue to the contractor must be in writing. An instruction need not be written on a specially printed form headed 'Architect's Instruction', although such forms are useful for collecting all instructions in one place. Even when they are used, a careful look through the files will normally unearth an instruction buried in the middle of a letter. A letter is quite acceptable as an instruction, so is a handwritten instruction given on site, provided it is signed and dated. Instructions contained in site-meeting minutes are valid if you produce

The architect's authority and duties

the minutes at a subsequent meeting and they are recorded as agreed. The position with regard to drawings is less certain. A drawing issued with a letter referring to its use on site is certainly an instruction, but a drawing issued with a compliments slip may be an instruction or it may be simply sent for comment. The contractor should make sure before carrying out the work shown thereon. If you simply send a copy of the employer's letter requesting that something be done, under cover of a compliments slip, it is not an instruction but merely an invitation to the contractor to carry out the work at his own cost.

There is no provision for oral instructions. If you give an oral instruction, the contractor can disregard it with impunity. If you confirm an oral instruction in writing, the instruction becomes effective only when the contractor receives the written confirmation. If the contractor attempts to confirm your oral instruction himself, this will probably be of no effect, even if you do not dissent, provided you do not acknowledge receipt. Oral instructions should be avoided.

Your authority to issue instructions is limited to those instructions which the conditions empower you to issue. A list of those instructions is given in Table 4.2 and they are discussed in more detail later. The flow chart Fig. 4.2 sets out the procedure.

The contractor must comply with your instructions forthwith. The exception is an instruction requiring a variation (addition, alteration, or omission) of any obligation or restriction imposed by the employer in the specification/schedules of work/bills of quantities in regard to access, limitation of working space, limitation of working hours, or the order of execution or completion of work (clause 3.6.2). The contractor need not comply with a clause 3.6.2 instruction if he makes a reasonable objection in writing. Any dispute as to the reasonableness of his objection must be referred to arbitration.

The contractor is entitled to require you to specify in writing the clause which empowers your instruction. You must respond immediately. It is good practice to specify the empowering clause in the instruction itself. When the contractor receives your response, he may do one of two things. He may carry out the instruction, in which case the clause you nominated will be deemed to be the empowering clause (whether it is or not) for all the purposes of the contract. Or he may give the employer a written notice to concur in the appointment of an arbitrator to decide if the clause you nominated does indeed empower the issue of your instruction. (The employer may give notice if he wishes, provided he does so before the contractor has complied with the instruction). It is thought that the contractor has the right to await the outcome of any arbitration before complying.

If the contractor does not carry out your instruction immediately, you may send him a written notice requiring him to comply with it (letter,

4.1 Authority

Table 4.2
Instructions empowered by IFC84

Clause	Instruction
1.4	Correcting inconsistencies between contract documents and drawings Correcting errors in the contract documents Correcting errors in particulars of a named person Correcting departures from the method of preparation of the contract bills
2.10	Not to make good defects
3.3.1	In regard to named persons as subcontractors: a) to change particulars to remove impediment b) to omit work c) to omit work and substitute a provisional sum Naming a person other than the person in the specification/schedules of work/contract bills
3.3.3	As necessary after determination of employment of named person: a) to name another person b) to require the contractor to make his own arrangements c) to omit the work
3.5.1	General power to issue instructions empowered by the conditions
3.6	Requiring a variation
3.8	To expend provisional sums
3.9	That setting out errors shall not be amended and an appropriate deduction be made from the contract sum
3.12	Requiring opening up or testing
3.13.1	Requiring opening up or testing at contractor's cost if similar work or materials have failed
3.14	To remove defective work from the site
3.15	Postponing work
6.3C.3(b)	To remove and dispose of any debris
7.4(b)	To remove temporary buildings, etc, after determination

The architect's authority and duties

Fig 4.2
Architect's instructions (clause 3.5)

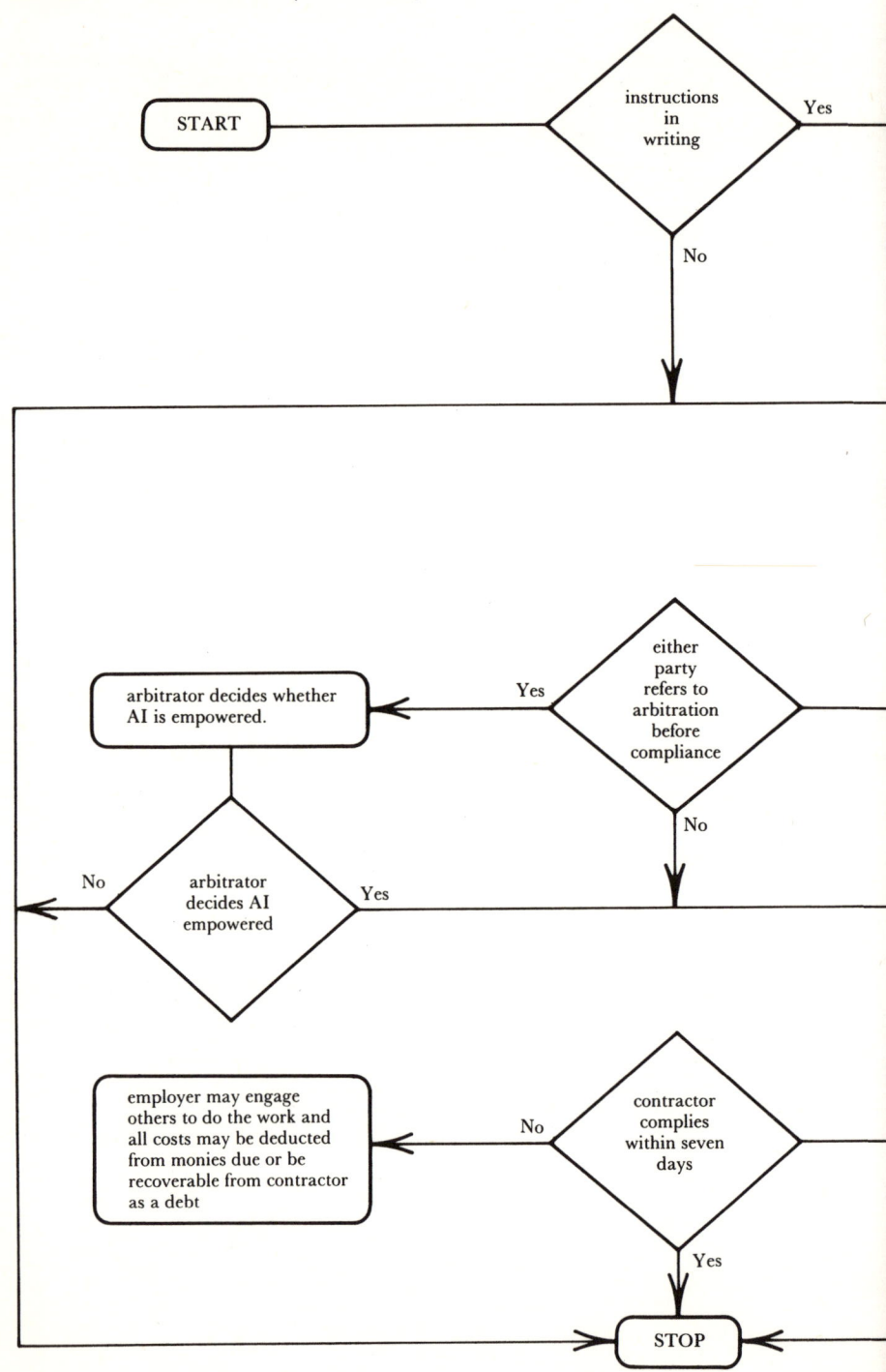

- START
- instructions in writing?
  - Yes → (continues)
  - No ↓
- either party refers to arbitration before compliance?
  - Yes → arbitrator decides whether AI is empowered.
    - arbitrator decides AI empowered?
      - No → (loop back)
      - Yes → (continues)
  - No ↓
- contractor complies within seven days?
  - No → employer may engage others to do the work and all costs may be deducted from monies due or be recoverable from contractor as a debt
  - Yes → STOP

46

4.1 Authority

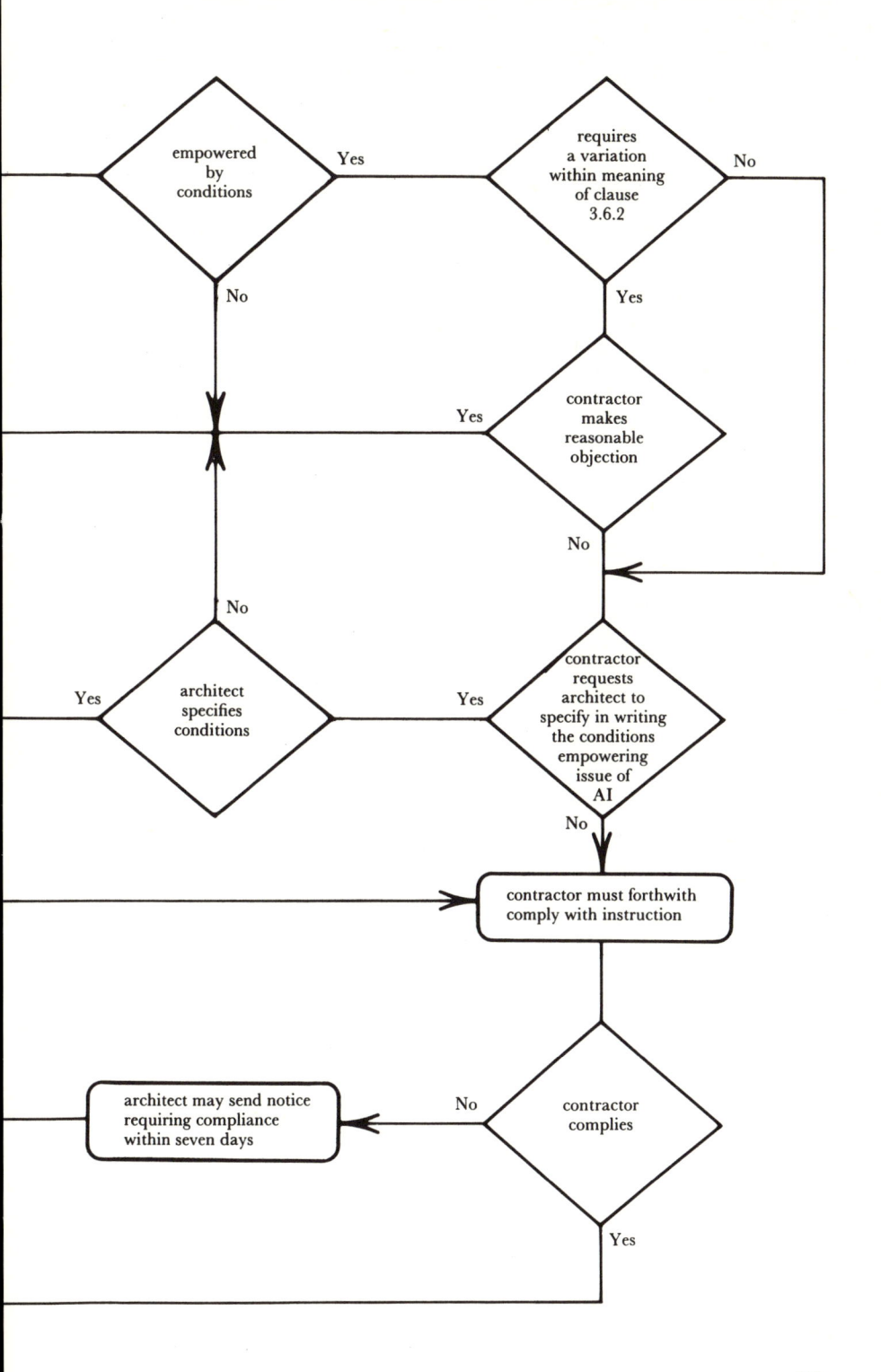

```
                                    requires
                                  a variation
        empowered      Yes       within meaning        No
           by                       of clause
        conditions                     3.6.2

              No                         Yes

                                     contractor
                          Yes           makes
                                      reasonable
                                      objection

                                         No

                                     contractor
                                      requests
          architect                   architect to
  Yes     specifies       Yes      specify in writing
         conditions                  the conditions
                                      empowering
                                       issue of
                                          AI
                                         No

                          →       contractor must forthwith
                                   comply with instruction

                                          ↓
    architect may send notice    No      contractor
     requiring compliance       ←         complies
       within seven days

                                         Yes
```

47

Fig 4.3). If he does not comply within seven days of receiving the notice, you should advise the employer that he may employ and pay others to do the work detailed in the instruction, including any work which it is necessary to carry out in order to comply with the instructions. Such additional work will usually consist of protective work, erecting scaffolding, cutting out, and reinstating. If the employer decides to take this course of action, he will expect you to handle the details. Wherever possible, you should obtain competitive quotations so that you will be able to show, if it becomes necessary, that you have had the work done at the lowest price it was reasonable to accept in the circumstances. When the work is completed, the employer has the right to deduct all the additional costs which he has incurred from monies due or which will become due to the contractor (ie, from certificates). Alternatively, he can opt to recover the cost as a debt. Generally, it will be easier for the employer to deduct the cost from monies payable on certificates. When deducting such money, the contractor is entitled to a brief statement showing how the figure has been made up. Note that the amount deductible is the additional costs—ie the difference between what the work would have cost had the contractor carried out the instruction and what it did cost in fact. You are entitled to charge extra fees, and the employer may include them in his computation of the costs together with any additional incidental expenses caused by the contractor's non-compliance.

4.1.4 Instructions in detail

The contract empowers you to issue instructions in a wide variety of circumstances. In some of these circumstances, you have an obligation to issue an instruction. These are marked by the use of the word 'shall', meaning 'must'. You must be aware of the extent of your powers and duties in each case.

Clause 1.4: correcting inconsistencies between contract documents and drawings; correcting errors in the contract documents; correcting errors in particulars of a named person; correcting departures from the method of preparation of the contract bills

There are four important points to note:
— you have an obligation to issue instructions under this clause
— your obligation to instruct does not depend on any notice from the contractor, although obviously he will notify you for his own benefit
— no inconsistencies, errors, or departures such as those mentioned will vitiate (invalidate) the contract
— if your instruction changes the quality or quantity of the work

4.1 Authority

Fig 4.3
Architect to contractor, giving notice requiring compliance with instruction

REGISTERED POST OR RECORDED DELIVERY

Dear Sir

Take this as notice under clause 3.5.1 of the conditions of contract that I require you to comply with my instruction number [*insert number*] dated [*insert date*], a further copy of which is enclosed.

If within seven days of receipt of this notice you have not complied, [*the employer*] may employ and pay other persons to execute any work whatsoever which may be necessary to give effect to the instruction. All costs incurred thereby will be deducted from money due, or to become due, to you under the contract or will be recovered from you as a debt.

Yours faithfully

Copy: Employer
      Quantity surveyor

deemed included in the contract sum (clause 1.2), or changes employer's obligations or restrictions, a variation results.

The onus is fairly and squarely on you. If you or the quantity surveyor make any errors, the employer will have to pay. It matters not that they are discovered only at a late stage in the progress of the works. This clause is admirably clear and leaves scant room for any fudging of the issue.

Clause 2.10: not to make good defects

This clause is discussed in detail in section 9.3.4. The important point to remember is that the employer's consent must be obtained before you issue the instruction; the contractor has no right of objection.

Clause 3.3.1: in regard to named persons as subcontractors: (a) change particulars to remove impediment; (b) omit work; (c) omit work and substitute a provisional sum

This clause is discussed in detail in section 8.2.2. You have an obligation to issue the instruction under this clause, provided that you are satisfied that the particulars specified have prevented the contractor from entering into a subcontract.

Clause 3.3.3: as necessary after determination of employment of named person: (a) name another person; (b) require the contractor to make his own arrangements; (c) omit the work

This clause is discussed in detail in section 8.2.2. You have an obligation to issue an instruction under this clause, subject to receiving a written notice from the contractor stating the circumstances of the determination.

Clause 3.6: requiring a variation

This is a most important clause, giving you the power to require the contractor to carry out variations or to sanction in writing any variation carried out by the contractor without instruction. The clause states that no such instruction or sanction will vitiate the contract. That is superfluous, because no exercise of a right conferred by the contract can vitiate that same contract. The clause goes into some detail regarding what a variation means. In general, it means what you would assume that it means, namely the alteration or modification of the design or quality or quantity of the works shown on the drawings and described by, or referred to, in the specification or schedules of work or bills of quantities (clause 3.6.1). Difficulties may arise, if bills of quantities are not used, in deciding just what is included and of what quality and, therefore, what is and what is not a variation (see section 3.1.2). The clause proceeds to spell out in detail what kinds of situation are included: the alteration of kind or standard of materials to be used in the works; the addition, omission, or substitution of any work; the removal from site of any work carried out or materials intended for use except where they are not in accordance with the contract.

This is just a development of the initial statement and requires no further comment. Variation is also said to mean (clause 3.6.2) the addition, alteration, or omission of any obligations imposed by the employer in the specification or schedules of work or bills of quantities in regard to access to, or use of, any particular parts of the site or the whole site; limitations of working space; limitations of working hours; the order of execution or completion of work.

Note that you can add, alter, or omit only if the obligations or restrictions are present in the first place. You cannot, for example, impose a limitation on the hours of working unless some limitation is already contained in the contract documents. If a limitation of hours is present in the contract documents, it would appear that you have the power to limit hours even further. The exercise of your power under this clause, however, is subject to the contractor's right of reasonable objection.

Clause 3.8: to expend provisional sums

This is mandatory. You must instruct the contractor how you wish to deal with any provisional sums.

Clause 3.9: that setting out errors shall not be amended and an appropriate deduction be made from the contract sum

You are responsible for giving the contractor accurately dimensioned drawings and for determining any levels required. The contractor is responsible for setting out the works correctly—ie, in accordance with the information you give him. If he sets out incorrectly, he must amend any errors arising and stand the cost himself. You may instruct the contractor not to amend errors arising from inaccurate setting out, but you must obtain the employer's consent before you issue your instruction, and you must instruct that an appropriate deduction for the errors must be made from the contract sum.

The deduction of an appropriate sum will pose difficulties. Bad setting out could result in, for example, several rooms becoming a metre shorter than intended. The wording of the clause appears wide enough

The architect's authority and duties

to cover not only the value of work and materials omitted, but also the reduction in value of the rooms to the employer—however that is to be ascertained. If the result of bad setting out is to leave the employer with a building substantially larger than he requires, the contractor would not be entitled to any increase for the additional work and materials, but he would face the possibility of a reduction to represent additional costs to the employer (such as increased rates, cleaning charges, and running costs).

What if the setting out is so bad that the building encroaches on a neighbour's land? Presumably the deduction must take all the employer's costs into account, but the situation is likely to arise only if the errors are minor, since if they are anything more, demolition is indicated. The contractor is given no choice in the matter, but, faced with what he considers to be an unreasonable attitude on the part of the employer, he is likely to seek arbitration or litigation.

Clause 3.12: requiring opening up or testing

This clause empowers you to instruct the contractor to open up for your inspection any work which has been covered up, or to arrange for any testing of materials whether or not they are already built in. You will instruct the contractor to open up or test because you suspect that work or materials are defective. You may have no alternative, but remember that, if the work or materials are found to be in accordance with the contract, the cost (including the cost of making good) must be added to the contract sum unless provision is made for opening up or testing in the specification or schedules of work or bills of quantities. If the work or materials are found to be not in accordance with the contract, all the costs must be borne by the contractor.

Clause 3.13.1: requiring opening up or testing at contractor's cost if similar work or materials have failed

This clause gives you a useful power to check for possible defective work. The exercise of that power will give you further responsibilities. Before you may issue an instruction, you must have discovered work or materials which have failed to be in accordance with the contract. After your discovery, the contractor must write to you stating what he intends to do immediately to ensure that there is no similar failure in the work already carried out or materials already supplied. His proposals must be at no cost to the employer. The important point is that only similar failures are under consideration. For example, if you discover that, in one section of the work, wall ties have not been provided in sufficient numbers, you will be looking to the contractor to satisfy you that wall ties have been properly provided in other parts of the building. Thus,

4.1 Authority

the contractor will not be concerned, on that occasion, with possible failures of damp-proof courses or foundations. It is debatable whether the contractor must also satisfy you that, in our example, the other wall ties have been properly bedded if that was not the reason for failure. On balance, he probably has no such duty. In most cases, of course, the contractor's proposals will give you the opportunity of checking that other defects are not present—although not necessarily in the case of wall ties.

Since, in the course of a contract, there will be many instances of failures, large and small, it would be wise for you to make sure that the contractor is aware of those instances when you consider his proposals under this clause to be necessary (letter, Fig 4.4). But it must be stressed that the onus is on the contractor and that you have no contractual duty to make him aware. Your power to issue an instruction arises after discovery of failure if:

— you have not received the contractor's proposals within seven days of discovering the failure; or

— you are not satisfied with the action proposed by the contractor; or

— you cannot wait for the contractor's written proposals because of safety considerations or statutory obligations.

You may then issue an instruction which requires the contractor to open up for inspection or to arrange for testing any work or materials, whether built in or not. Your instruction should state that the work will be at no cost to the employer. The whole process is at the contractor's own cost, whether or not the opening up or testing discovers further failures (this is in contrast to the provisions of clause 3.12). The contractor must forthwith comply with your instruction. If he does not, you may obtain the employer's consent to applying clause 3.5.1 remedies (see section 4.1.3).

Clause 3.13.2 gives the contractor 10 days from receipt of your instruction to decide whether to object to compliance. His right of objection is stated to be without prejudice to his obligation to comply. This means that he must still comply even though objecting. If he decides to object, he must inform you in writing, stating his reasons. You have seven days from receipt of his objection to withdraw your instruction, or to modify your instruction to take care of the contractor's objection.

If you take neither action, any dispute regarding whether the nature or extent of the opening up or testing was reasonable in all the circumstances is referred automatically to arbitration. The arbitration is stated to be in accordance with article 5, including article 5.4, even if the appendix states that that article is not to apply (see section 13.5).

The arbitrator is to decide whether your instruction is fair and reasonable. If he decides that it is not, he must decide what amount the

Fig 4.4
Architect to contractor, following failure of work

Dear Sir

When I visited site today, I noted that [*specify work or materials*] failed to be in accordance with the contract.

In accordance with clause 3.13.1 of the conditions of contract, I require you to state in writing, within seven days of the date of this letter, what action you will immediately take at no cost to the employer to establish that there is no similar failure in work already executed/materials or goods already supplied [*omit as appropriate*].

Yours faithfully

employer must pay to the contractor for carrying out the work, including making good. It is clearly envisaged that the arbitrator may find that your instruction is fair in essence but that you have gone too far in requiring existing work to be opened up. In such a case, he has power to order the employer to pay a contribution. The contractor, however, must still comply with your instruction. In the majority of cases, the contractor is unlikely to wish to force arbitration, but the wording of the clause makes arbitration inevitable if you ignore any objection the contractor makes. The contractor may take advantage of that to try and force concessions when he makes objection. In practice, no matter how automatic arbitration is said to be, it will not take place against the wishes of both parties. It will be seen that, if all parties take the maximum allowable time to act under the various parts of the clause, some 26 days will have elapsed between the discovery of the failure and the date on which arbitration becomes automatic. The procedures and options in this clause are set out in the flowchart, Fig 4.5.

Clause 3.14: to remove defective work from the site

This clause empowers you to order that defective work or materials be removed from site. Defective work or materials is work or materials not in accordance with the contract. Note that you have no power to simply order defective work or materials to be rectified (except under clause 2.10, defects liability). To be effective, your instruction must order removal from site. In most cases, it is to be expected that the contractor will correct defective work or materials without the necessity of an instruction, but remember that you cannot put any sanction into operation until you have instructed removal and the contractor has not complied. So it is worth while sending an instruction under this clause whenever such work or materials comes to your attention.

Clause 3.15: postponing work

You are entitled to issue an instruction to postpone any of the work required by the contract. There is a price to pay and the employer has to pay it, so take care. If you postpone work, the contractor can claim an extension of time (clause 2.4.5) and loss and/or expense (clause 4.12.5). Under clause 7.5.3, he may determine his employment if the whole or substantially the whole of the works are suspended for a continuous period of one month due to postponement, among other things (see section 12.2.2). It has been held that, in certain circumstances, an instruction given on another matter may imply postponement with all its consequences. The situation should not arise if you take care to

The architect's authority and duties

Fig 4.5
Failure of work

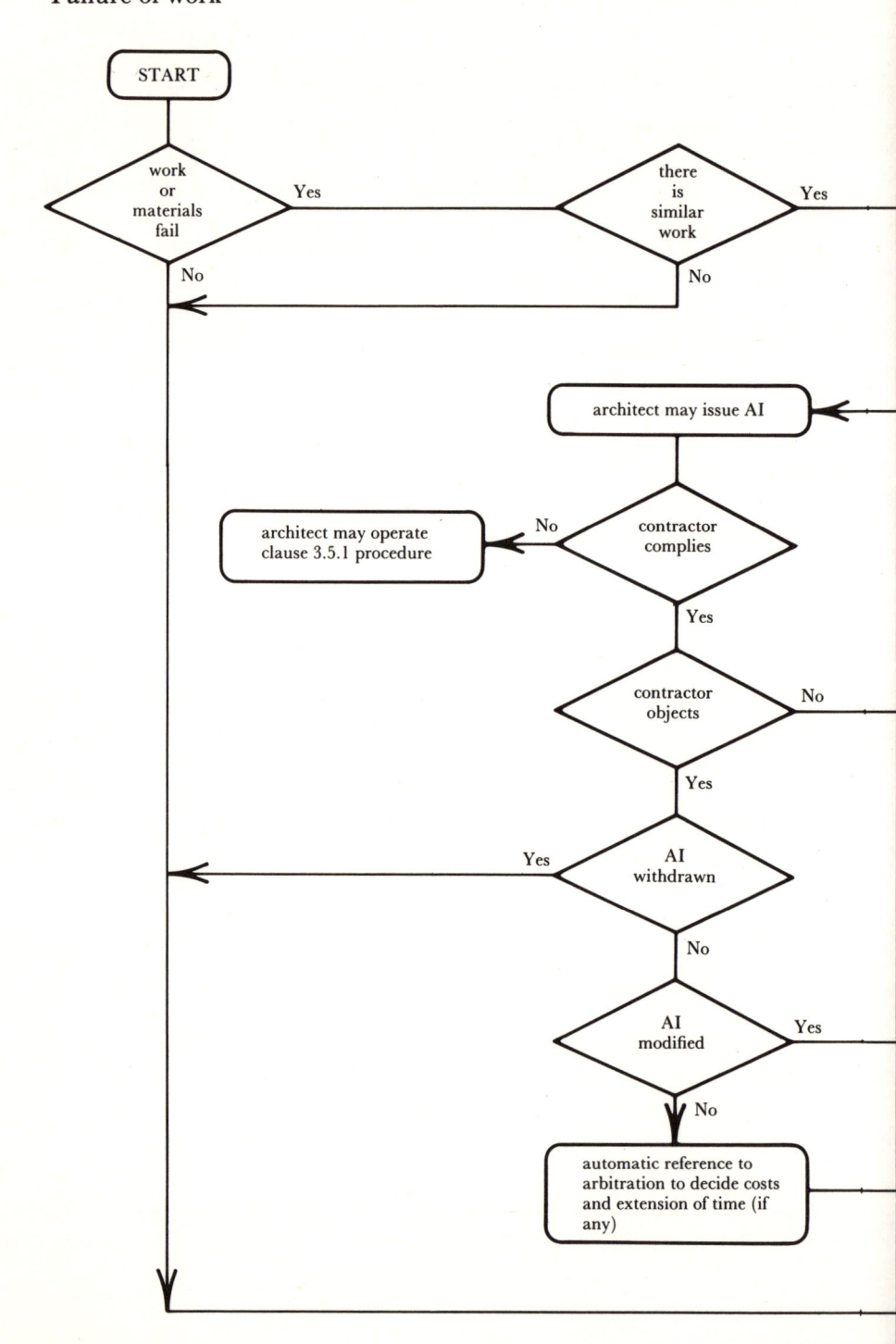

```
                    START
                      │
         ┌────────────┴────────────┐
    work or                   there is
   materials ──Yes──────────► similar ──Yes──►
     fail                       work
      │                          │
      No                         No
      │                          │
      │◄─────────────────────────┘
      │
      │                                    architect may issue AI ◄──
      │                                              │
      │            architect may operate      contractor
      │            clause 3.5.1 procedure ◄─No─ complies
      │                                              │
      │                                             Yes
      │                                              │
      │                                         contractor
      │                                          objects ──No──►
      │                                              │
      │                                             Yes
      │                                              │
      │◄──────────────────Yes───────────────── AI withdrawn
      │                                              │
      │                                             No
      │                                              │
      │                                            AI
      │                                          modified ──Yes──►
      │                                              │
      │                                             No
      │                                              ▼
      │                                    automatic reference to
      │                                    arbitration to decide costs
      │                                    and extension of time (if
      │                                    any)
      ▼
```

4.1 Authority

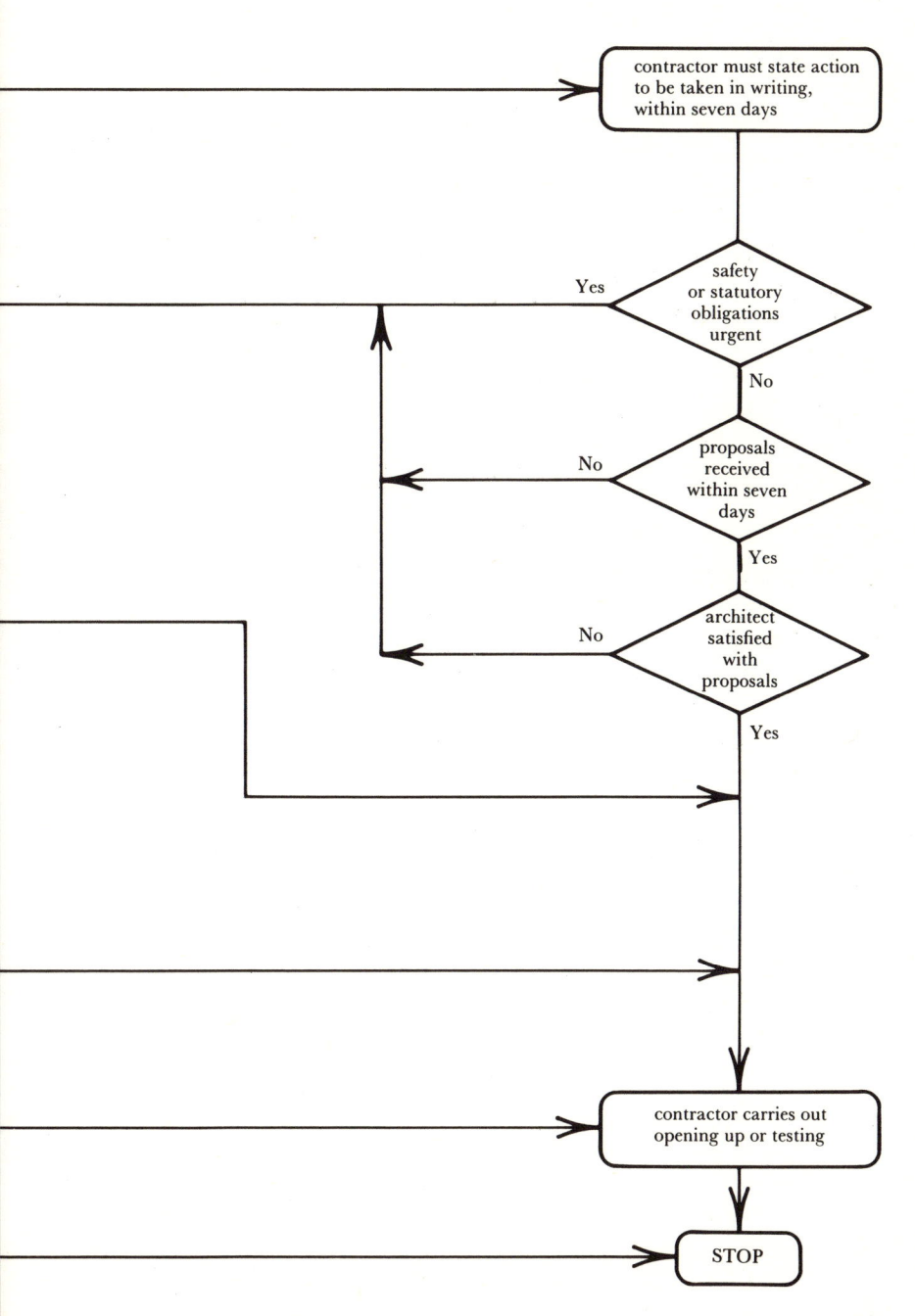

```
                                    ┌─────────────────────┐
                                    │ contractor must state action │
────────────────────────────────────▶│ to be taken in writing,    │
                                    │ within seven days           │
                                    └──────────┬──────────┘
                                               │
                                               ▼
                          Yes            ╱ safety ╲
          ◀────────────────────────────╱ or statutory ╲
          │                            ╲ obligations ╱
          │                             ╲  urgent  ╱
          │                                  │ No
          │                                  ▼
          │                          ╱ proposals ╲
          │         No              ╱  received   ╲
          ◀─────────────────────────╲ within seven ╱
          │                          ╲   days    ╱
          │                                │ Yes
          │                                ▼
          │                          ╱ architect ╲
          │         No              ╱  satisfied  ╲
 ─────────┼─────────────────────────╲    with    ╱
          │                          ╲ proposals ╱
          │                                │ Yes
          │                                ▼
 ─────────┼────────────────────────────▶
          │                                │
          │                                ▼
          │                    ┌───────────────────────┐
 ─────────┼───────────────────▶│ contractor carries out │
                                │  opening up or testing │
                                └──────────┬────────────┘
                                           ▼
 ──────────────────────────────────▶ ┌──────┐
                                     │ STOP │
                                     └──────┘
```

quote the correct empowering clause in each case and you use the wording of the clause as far as appropriate.

Clause 6.3C.3(b): to remove and dispose of any debris

If loss or damage is caused to the works by clause 6.3 perils and the works are alterations or extensions to an existing building for which the employer has taken out insurance, you are empowered to issue instructions requiring the contractor to remove and dispose of any debris. There is a proviso that no notice of determination shall have been served by either party or, if a notice has been served, that the arbitrator has decided against it.

Clause 7.4(b): to remove temporary buildings, etc, after determination

Although not prefaced by the mandatory 'shall', this is an instruction which you are obliged to issue at some time after the employer has determined the contractor's employment. For a fuller discussion, see section 11.1.7.

4.2 Duties

4.2.1 Duties under the contract

In section 4.1.1, your duties under the contract were seen to flow directly from your contract with the employer. Many of those duties can be recognised because they are preceded by the word 'shall'—eg, 'the architect shall . . .'. A full list is contained in Table 4.1.

In performing your duties, you are expected to act competently; as an architect, you will be expected to act with the same degree of skill and care as the average competent architect. If you profess greater than average skill, either generally or in some special area, that is the standard by which you will be judged.

Clause 1.1, Contractor's obligations, paradoxically places a potentially heavy duty on you. In essence, it states that if approval of workmanship or materials is a matter for your opinion, the quality and standards must be to your reasonable satisfaction. At first sight, this clause appears to give you considerable power, and so it does. The catch is contained in clause 4.7, which states that the issue of the final certificate is conclusive evidence that where anything is reserved to your approval, it is to your reasonable satisfaction. Architects commonly specify that various items are to be to their approval, sometimes even going so far as to state generally that 'unless otherwise stated, all workmanship and materials are to be the approval of the

4.2 Duties

architect'. The combined effect of clauses 1.1 and 4.7 is that, if you reserve anything to your approval, the final certificate confirms that you do approve it. That is the case whether or not you have specifically expressed approval during the course of the work. Therefore, if you put a general-approval clause in the contract documents, you place on yourself a duty of approving everything to which it applies. If you miss anything, the final certificate will make it deemed approved. Moreover, if you approve something which is not in accordance with the contract, your approval will override the contract requirements. You cannot say later that it was not in accordance with the contract and therefore unacceptable. You will be wise severely to limit the items to which you reserve the right of approval. Table 4.3 lists the certificates you are to issue.

Table 4.3
Certificates to be issued by the architect under IFC84

Clause	Certificate
2.6	Certificate of non-completion
2.9	Certificate of practical completion
2.10	Certificate of making good defects
4.2	Interim certificates
4.3	Interim certificate on practical completion
4.6	Final certificate

4.2.2 General duties

This is not the place to discuss in detail your general duties to your client and to third parties, but one aspect of your general duties affects the contract. You have a duty to your client to be familiar with those parts of the law which affect your work. For example, your client will expect you to have a thorough knowledge of the planning laws. This is not the specialist knowledge expected of a lawyer who deals with nothing but planning appeals, but is the knowledge which you require to advise your client and make successful planning applications on his behalf. Similarly, you are expected to have a thorough knowledge of the various forms of contract, so that you can advise your client which contract is most appropriate for a particular job. You must be knowledgeable about the particular contract which you advise him to use.

You must be aware of the pitfalls in a contract and the correct interpretation of each clause as shown by any applicable case law. If you have never read a contract in your life, much less a legal commentary,

The architect's authority and duties

then you will be unable to carry out your duty under the contract; you will be unable to act fairly within the meaning of the contract; you will be unable to advise your client when a particularly difficult contractual point arises; and you will make more mistakes than the average competent architect.

If your client is put to unnecessary expense by your inadequate knowledge of contractual provisions, he may well sue.

4.3 Summary

Authority

— your authority depends on what you have agreed with the employer
— only the employer can take action against you in contract
— employer and contractor can take action against you in tort
— you have a dual role once the contract is signed
— you are not a quasi-arbitrator; you are not immune from actions for negligence for your decisions
— the contractor is entitled to see your authority only in terms of the contract
— only the architect named in the contract or his authorised representative can exercise any powers under the contract
— if you take over a contract from another architect, you cannot alter his decisions with which you disagree, but you should inform the employer
— matters reserved for you to decide cannot be delegated to another (eg, the quantity surveyor)
— instructions must be clear and in writing
— you may issue only those instructions which the contract specifically empowers you to issue
— you must be aware of your specific powers and duties in the case of each instruction which you are empowered to issue

Duties

— you must be as skilled as the average architect
— you will be judged on any skill which you profess to possess over and above the average
— if items are stated to be to your approval, you have a duty to approve each one
— you have a duty to know the law as it affects you in the performance of your profession
— you must know the law as it affects the contract

5 The contractor's obligations

5.1 Express and implied obligations

5.1.1 Legal principles

Apart from the express terms of the contract, the general law requires that the contractor will do three things:
— the contractor must carry out his work in a good and workmanlike manner—ie, show the same degree of competence as the average contractor experienced in carrying out that type of work.
— the contractor will supply good and proper materials.
— the contractor undertakes that the building will be reasonably fit for its intended purpose.

These obligations may be modified by the terms of the contract itself. Also under the general law, the contractor is responsible to the employer for the work done, and goods and materials provided, by his subcontractors and, in common with most standard form building contracts, under IFC84 the contractor is responsible to the employer for all the defaults of a subcontractor, whether named or otherwise. But in the case of named subcontractors (see section 8.2.2), this liability is substantially modified by the very wide terms of clause 3.3.7, which exempts the contractor from responsibility to the employer for design and allied failures in the named subcontractor's work. This apart, the position under IFC84 is that the main contractor is liable to the employer for all other subcontractors' defaults, of fabrication, workmanship, or otherwise.

The contractor's obligations

Statutory obligations are also imposed on contractors by the Supply of Goods and Services Act 1982, although these statutory obligations are included in the express terms of IFC84.

The contract must be read against this background. It imposes many specific duties on the contractor, some of which alter or affect the common-law position. These obligations are scattered throughout the printed form, and only the more important of them are collected in this chapter.

Table 5.1 summarises the contractor's powers and duties under the express terms of the contract.

5.1.2 Execution of the works

Clause 1.1 requires the contractor to 'carry out and complete the works in accordance with the contract documents', which are specified in the second recital; this is a basic and absolute obligation. It is not qualified in any way. The contractor must bring the works to a state where they are 'practically completed' so that you may issue your certificate under clause 2.9. This is what the contractor must do, no matter what difficulties he may encounter, but subject to what is said in clause 7.8 about the determination of his employment under the contract for causes outside the control of either party.

The work must be carried out and completed 'in accordance with' the contract documents as defined. You must make sure that the description of the work is adequate, and care should be taken to use precise wording. Generalisations are impossible to enforce. The contract documents must contain all the requirements which the employer wishes to impose, and you should avoid the use of phrases such as 'of good quality' or 'of a durable standard'.

The contractor must complete all the work shown in, described by, or referred to in the contract documents. His obligation is ended only when you issue your certificate of practical completion. Thereafter he must remedy defective work during and immediately after the specified defects liability period (see section 9.3).

The proviso to clause 1.1 states that if approval of workmanship or materials is a matter for your opinion then the quality and standard must be to your reasonable satisfaction. The effect of this is discussed in section 4.2.1.

The basic contractual obligation is amplified by clause 2.1. Once he is given possession of the site, the contractor must begin and proceed 'regularly and diligently' with the works and complete them on or before the specified completion date, as extended. Failure to proceed 'regularly and diligently' is one of the grounds which may give rise to determination of his employment under the contract by the employer (see clause 7.1 (b)).

5.1 Express and implied obligations

Table 5.1
Contractor's powers and duties under IFC84

Clause	Power/duty	Precondition/comment
1.1.	**Duty** Carry out and complete the works in accordance with the contract documents	
1.6	**Power** Inspect the contract documents	At reasonable times: ie, during the employer's normal hours of business
1.8	**Duty** Use the specification/bills/schedules/drawings/details only for the purpose of the contract	
1.11.	**Duty** Use off-site goods and materials which have been paid for by the employer only for the works and not otherwise to remove them or permit their removal	
2.1	**Duty** Begin the works when given possession of the site	
2.1	**Duty** Regularly and diligently proceed with the works and complete them on or before the date for completion specified in the appendix	This is subject to the provisions for extension of time in clause 2.3
2.3	**Duty** Notify the architect in writing forthwith of any cause of delay	If it becomes reasonably apparent that progress of the works is being or is likely to be delayed. The duty is in respect of any cause of delay and is not confined to the events specified in clause 2.4. The second part of this duty does not require the contractor to expend money
	Duty Constantly use his best endeavours to prevent delay and do all that may reasonably be required to the architect's satisfaction to proceed with the works	
	Duty Provide the architect with sufficient information to enable him properly to exercise his duties as regards extensions of time	This is conditional on a request from the architect, and the information must be 'reasonably necessary'

The contractor's obligations

Table 5.1
Contractor's powers and duties under IFC84—*cont.*

Clause	Power/duty	Precondition/comment
2.4.7	**Duty** Make specific written application to the architect for any necessary instructions, drawings, details, or levels	Failure to do so at the right time vitiates any claim for an extension of time on the ground of late instruction
2.7	**Duty** Pay or allow to the employer liquidated damages at the rate specified in the appendix	If the works are not completed by the specified or extended date for completion *and* If the architect has issued a certificate of non-completion under clause 2.6 *and* If the employer has required liquidated damages in writing not later than date of final certificate for payment
2.10	**Duty** Make good any defects, shrinkages, or other faults at no cost to the employer	If the defects etc appear and are notified to the contractor by the architect not later than 14 days after the expiry of the defects liability period *and* If they are due to materials or workmanship not in accordance with the contract or to frost occuring before practical completion *and* If the architect has not instructed otherwise
3.1	**Power** Assign the contract	If the employer gives consent in writing
3.2	**Power** Subcontract any part of the works	If the architect consents in writing and subject to clause 3.3 below There are conditions which must be imposed in any ensuing subcontract
3.3.1	**Duty** Enter into a subcontract using section III of NAM/T with any named person, not later than 21 days of entering into the main contract	The person must be named in the specification/schedule of work/bills

5.1 Express and implied obligations

	Duty Immediately inform the architect if unable so to enter into a subcontract in accordance with the particulars given in the contract documents and specify which particulars have prevented the execution of the subcontract **Duty** Notify the architect of the date of entering into a subcontract with a named person	
3.3.2(c)	**Power** Make reasonable objection to entering into subcontract with a named person	Must be made with 14 days of the architect's issuing a clause 3.8 instruction
3.3.3	**Duty** Advise the architect as soon as is reasonably practicable of any events which are likely to lead to the determination of the named person's employment under a subcontract **Duty** Notify the architect in writing if the named person's employment is determined before completion of the subcontract work, stating the circumstances	Whether or not the architect has already been advised of event likely to lead to determination
3.3.6(b)	**Duty** Take such reasonable action as is necessary to recover from the named subcontractor any additional amount payable as a result of default or failure **Duty** Account to the employer for amounts so recovered **Duty** Repay to the employer any additional amounts involved	Recovery is under clause 27.3.3 of the subcontract NAM/SC The contractor is not required to commence arbitration proceedings or litigation unless the employer indemnifies him against legal costs Only to the extent he has failed in his clause 3.3.6 duty
3.4	**Duty** Keep a competent person in charge of the works at all reasonable times	
3.5.1	**Duty** Forthwith carry out all written instructions issued by the architect	Provided the instruction is one which the contract empowers the architect to issue. The contractor has a right to make reasonable objection in writing to compliance with an instruction which modifies any obligations or restrictions imposed by the employer in the specification/schedule/bills about site access, limitations of working space/hours, or execution/completion of work in a specific order

65

The contractor's obligations

Table 5.1
Contractor's powers and duties under IFC84—*cont.*

Clause	Power/duty	Precondition/comment
3.5.2	**Power** Request the architect to specify in writing which contract clause empowers the issue of an instruction	When he receives an instruction
	Power Serve written request on employer to concur in the appointment of an arbitrator	If dissatisfied with architect's reply and not willing to comply
3.7	**Power** Agree with employer the adjustment to the contract sum in respect of variation instructions and of instructions on the expenditure of a provisional sum	Prior to contractor's complying with the instruction
3.9	**Duty** Set out the works accurately	Architect determines levels and provides contractor with accurately dimensioned drawings to enable this to be done
	Duty Amend at his own cost any errors arising from inaccurate setting out	The architect may instruct otherwise with employer's consent
3.11	**Duty** Permit the execution of work not forming part of the contract to be carried out by the employer or person employed or engaged by him concurrent with the contract works	If contract documents so provide
	Power Consent to the carrying out of such work by others	Where employer requests and contract documents do not so provide Consent must not be unreasonably withheld
3.12	**Duty** Bear cost of opening up and testing and consequential costs of making good	If inspection and test shows that materials, goods, or work are not in accordance with the contract
3.13.1	**Duty** State in writing to the architect the action which the contractor proposes to take immediately to establish that there is no similar failure of work, etc	Where such failure is discovered during the carrying out of the works
	Duty Forthwith comply with any architect's instruction requiring opening up for inspection and testing	Unless within 10 days of receipt of the instruction the contractor objects to compliance, stating his reasons in writing. Then, if within seven days of receipt of the contractor's objection the architect does not withdraw or modify his instructions in writing, the dispute or difference is referred to arbitration

5.1 Express and implied obligations

4.5	**Duty** Before, or within a reasonable time after, practical completion send to the architect (or to the QS if the architect so instructs) all documents reasonably required for the purposes of the adjustment of the contract sum	
4.11	**Power** Make written application to the architect within a reasonable time	If it becomes apparent that regular progress is being materially affected by one or more of the specified matters or due to deferment of possession of the site by the employer
	Duty Submit to the architect or quantity surveyor such information as is reasonably necessary to enable an ascertainment of direct loss and/or expense to be made	
4.12.1	**Duty** Make specific written application to the architect for any necessary instructions, drawings, details or levels	Failure to do so at the right time vitiates any claim for loss and/or expense on the ground of late instruction
5.1	**Duty** Comply with, and give all notices required by any statute, statutory instrument, rule, order, regulation, or byelaw	As applicable to the works
	Duty Pay all fees and charges in respect of the works	The amount of such fees, etc, is added to the contract sum unless they are required by the specification/schedules of work/contract bills to be included in the contract sum
5.2	**Duty** Immediately give to the architect a written notice specifying any divergence between the statutory requirements and contract documents, or between such requirements and any architect's instruction	If the contractor finds any divergence
5.4.1	**Duty** Supply such limited materials and execute such limited work as are reasonably necessary to secure immediate compliance with statutory requirements	In an emergency (eg, a dangerous structure notice) and if it is necessary to do this prior to receipt of an instruction from the architect
5.4.2	**Duty** Forthwith inform the architect of such emergency compliance	

The contractor's obligations

Table 5.1
Contractor's powers and duties under IFC84—*cont.*

Clause	Power/duty	Precondition/comment
5.7 (and supplemental condition E)	**Duty** Comply with the fair wages provisions	If the employer is a local authority
6.1	**Duty** Indemnify the employer against any expense, liability, loss, claim, or proceedings whatsoever in respect of personal injury to or death of any person	The claim must arise out of, or in the course of, or be caused by, the carrying out of the works *and* not be due to any act or neglect of the employer or any person for whom he is responsible
	Duty Similarly indemnify the employer against property damage	If the claim arises out of, or in the course of, or is caused by, the carrying out of the works *and* is due to any negligence, ommission, or default of the contractor, his servants or agents or that of any subcontractor, his servants or agents
		The indemnity does not extend to loss or damage which is at the sole risk of the employer under clause 6.3B or 6.3C (if applicable)
6.2.1	**Duty** Maintain, and cause any subcontractor to maintain, necessary insurances in respect of injury to persons or property	The obligation to maintain insurance is without prejudice to the contractor's liability to indemnify the employer
6.2.2	**Duty** Produce, and cause any subcontractor to produce, for the architect's inspection documentary evidence of insurance cover	When reasonably required to do so by the architect, who may (but not unreasonably or vexatiously) require production of the policy (policies) and premium receipts
6.2.4	**Duty** Maintain joint names of employer and contractor for such amounts of indemnity as are specified in the contract documents for insurance against damage to property, other than the works, caused by collapse, subsidence, etc	A provisional sum must be included in the contract documents
6.2.4	**Duty** Deposit with the architect the policy (policies) and premium receipts	
6.3A	**Duty** Insure the works against clause 6.3 perils for their full reinstatement value	New buildings: the obligation continues until the date of issue of the certificates of practical completion. Clause 6.3A.3 enables this cover to be by means of the contractor's 'all risks' policy
	Duty Deposit with the architect the policy (policies) and premium receipt	
	Duty With due diligence, restore work damaged, replace or repair any unfixed goods or materials which have been destroyed or damaged, remove and dispose of debris, and proceed with the carrying out and completion of the works	On acceptance by the insurers of any claim under clause 6.3A insurance

5.1 Express and implied obligations

6.3B	**Power** Require the employer to produce for inspection the insurance policy and last premium receipt **Power** Insure in joint names all work executed, etc, against loss or damage by clause 6.3 perils	Only where the employer is not a local authority If the employer fails to produce receipt on request
6.3B	**Duty** Forthwith notify the architect *and* the employer of the extent, nature, and location of any loss or damage affecting the works or any unfixed materials or goods **Duty** With due diligence, restore work damaged, replace or repair any unfixed goods or materials which have been destroyed or damaged, remove and dispose of debris, and proceed with the carrying out and completion of the works	The contractor must do this on discovering the loss or damage. Clause 6.3B covers new buildings where the employer has accepted the insurance risk
6.3C	**Power** Request employer to produce receipt showing that he has an effective policy against loss or damage by clause 6.3 perils **Power** Insure existing structures in joint names and for that purpose enter on the premises to make a survey and inventory **Duty** Forthwith give written notice to the architect and the employer of the extent, nature, and location of any loss or damage occasioned by clause 6.3 perils **Power** Serve notice determining his employment **Power** Request the employer in writing to concur in the appointment of an arbitrator **Duty** With all due diligence, reinstate and make good loss or damage and proceed with the carrying out and completion of the work	In existing structures Where such risk is at sole risk of employer If employer fails to produce premium receipt when requested Upon discovering the loss or damage If just and equitable so to do If the employer serves notice determining the contractor's employment *and* the contractor alleges that it is not just and equitable so to do. If no notice of determination is served or if the arbitrator decides against the notice of determination
7.4	**Duty** Give up possession of the site of the works **Duty** Remove from the works any temporary buildings, plant, tools, equipment, goods, and materials belonging to, or hired by, the contractor	In the event of determination by the employer under clauses 7.1, 7.2, or 7.3 As and when so instructed in writing by the architect
7.5	**Power** Serve a default notice on the employer by registered post or recorded delivery, specifying the default alleged	If the employer: Does not pay an amount properly due to the contractor on any interim or final certificate; *or* interferes with or obstructs the issue of any certificate *or* the carrying out of the whole or substantially the whole of the uncompleted works (other than the execution of *(continued)*

The contractor's obligations

Table 5.1
Contractor's powers and duties under IFC84—*cont.*

Clause	Power/duty	Precondition/comment
		remedial works [defects liability]) is suspended for a continuous period of one month by reason of —Architect's instructions under 1.4. (Inconsistencies); 3.6 (Variation) 3.15 (Postponement) unless caused by contractor's neglect or default —Contractor not having received in due time necessary instructions etc from the architect for which he made specific written application at the right time —Delay in execution of work by employer himself or by others engaged or employed by him or failure to execute such work, or delay or failure in supply of materials or goods which the employer undertook so to supply —Failure by employer to give in due time ingress to or egress from the site if so agreed.
	Power Determine his employment under the contract by written notice served on the employer by registered post or recorded delivery	If the employer continues his default for 14 days after receipt of the default notice or thereafter repeats the same default The notice must not be given unreasonably or vexatiously
7.6	**Power** Determine his employment under the contract by written notice served on the employer by registered post or recorded delivery	If the employer becomes bankrupt or is insolvent, etc, or has a receiver appointed
7.7	**Duty** Remove from site with all reasonable despatch all his temporary buildings, plant, tools, equipment, goods, and materials, and give facilities to his subcontractor to do the same	Where the contractor determines his own employment
7.8	**Power** Forthwith determine his employment under the contract by written notice served on the employer by registered post or recorded delivery	If the carrying out of the whole or substantially the whole of the uncompleted works (except work required under 2.10 defects liability) is suspended by reason of *force majeure* or loss or damage to the works occasioned by clause 6.3 perils (unless caused by the negligence of the contractor or those for whom he is vicariously responsible) or civil commotion
7.9	**Duty** Remove from site with all reasonable dispatch all his temporary buildings, etc, and give facilities to his subcontractors to do the same	

5.1 Express and implied obligations

The trouble is that there is no generally accepted definition of 'regularly and diligently' in this context, and the judges have not been able to come up with a clear-cut answer. According to one line of authority, merely 'going slow' is not a breach of contract as such. It becomes a breach of contract only if there is ultimately a delay in completion. It is a question of fact whether the contractor is going ahead 'regularly and diligently', and this is clearly to be judged by the standards to be expected of the average competent and experienced contractor.

If, sensibly, you have required the contractor to provide you with a programme, even though this is not a contract document, it is a standard against which the contractor's progress can be measured, but you should beware of approving any programme submitted by the contractor lest he try to use that as a lever against you if things do not go as planned.

5.1.3 Workmanship and materials

The contract is strangely silent about the standards of workmanship and materials or even the skill and care to be expected of the contractor in the performance of his obligations. These matters must be deduced from the other contract documents and under the general law.

Certainly, he is expected to show a reasonable degree of competence and to employ skilled tradesmen and others, although you have no power to direct how he should carry out his work or to require him to replace employees who are, in your view, unsatisfactory.

The quality and quantity of work must be adequately defined in the contract documents, as is made plain by clause 1.2, the provisions of which are important in defining the contractor's obligations. In general, it may be said that any extra cost which results from faulty description in the contract documents falls on the employer. For example, if there is an inconsistency in the contract documents or an error of description, you will need to issue the apropriate instruction under clause 1.4.

Clause 1.2 should be studied carefully. Unlike JCT80, IFC84 makes no provision in the contract that the contractor is to provide materials, goods, and workmanship 'so far as procurable', which would be a valuable protection to him.

The other contractual references to materials and goods are in clauses 1.10 and 1.11, which are concerned solely with the transfer of ownership in materials and goods. You should not be misled by the statement in those provisions that 'such materials and goods shall become the property of the employer', as this is not necessarily the case. If the contractor is not the legal owner of the goods, he cannot pass

The contractor's obligations

title to the employer, and so you need to take great care before including the value of unfixed goods or materials or off-site items in any interim certificate unless the contractor provides you with proof of ownership (eg, a copy of his sale contract from the supplier).

5.1.4 Statutory obligations

Clause 5.1 places on the contractor an obligation to comply with all statutory obligations and to pay all fees and charges in respect of the works which are legally recoverable from him (eg, fees under the building regulations). Clause 5.2 also imposes on him an obligation to give you immediate written notice if he discovers any divergence between the contract documents or one of your instructions and the statutory requirements.

Contractually, clause 5.3 exempts the contractor from liability to the employer if the works do not comply with statutory requirements provided he has carried the work out in accordance with the contract documents or any instruction of yours where, for example, he does not spot the divergence. The wording in clause 5.2 is, 'if the contractor finds any divergence . . .', and, unless he does so, he is under no obligation to notify you. However, whatever the position may be as between contractor and employer under the contract, the exempting provision cannot exempt him from his duty to comply with the building regulations and other statutory obligations, liability under which may be absolute. But the wording is sufficiently wide to protect the contractor from any action by the employer, which leaves you in the firing line if the fault is yours.

So far as fees and charges in connection with the works are concerned, these are to be added to the contract sum so that the contractor is reimbursed, although clause 5.1 establishes that the contract documents can require such fees and charges to be included in the tender sum. The reference to 'rates and taxes' is interesting and is designed to cover those (comparatively rare) occasions when site huts and so on are ratable.

5.1.5 Person-in-charge

Clause 3.4 requires the contractor to keep on the works 'a competent person-in-charge', and he must do this at all reasonable times—ie, during normal working hours. This person is intended to be the contractor's full-time representative on site, but his appointment and replacement are not subject to your approval.

'Competent' can only mean what it says—ie, having sufficient skill and knowledge—and the person-in-charge is the contractor's agent for the purpose of accepting your written instructions, which are then *deemed* to have been issued to the contractor. If you wish to ensure that the person-in-charge is subject to your approval, this is a matter which can be dealt with in the contract documents.

5.1.6 Levels and setting out

Although it is your duty to determine any levels which may be required for the execution of the works and to provide the contractor with accurately dimensioned drawings to enable him to set out, clause 3.9 obliges the contractor accurately to set out the work in accordance with your instructions. He is made responsible for his own setting-out errors and must amend them at his own cost.

The sensible interpretation of the last sentence of clause 3.9 is that you may, with the employer's consent, instruct the contractor not to amend setting-out errors and make an appropriate adjustment to the contract sum, though exactly how this is to be assessed is not stated.

5.2 Other obligations

5.2.1 Access to the works and premises

Unlike JCT80, IFC84 does not expressly provide that you or your representative should have access to the works at all reasonable times. It is unnecessary for that right to be referred to expressly, because it is implied under the general law. But IFC84 does contain a gap, because you also need access to the workshops or other places of the contractor where work is being prepared for the contract. We suggest, therefore, that where appropriate you make a special provision to this effect in the contract documents, following the wording of JCT80, clause 9, because it is probable that the general law would not give you that right of access.

5.2.2 Drawings, details, and information

Clause 1.7 requires you to provide the contractor from time to time with further drawings, details, and information enabling him to carry out and complete the works, and failure to do this is a breach of contract for which the employer would in principle be liable in damages. But the contractor is also under contractual duty to make a written request for particular details or instructions. This is to be deduced from clause 2.4.7, which states that the contractor must have made a *specific* written application to you for any necessary instructions, drawings, details, or levels, 'on a date which having regard to the date for completion stated in the appendix or any extended time then fixed was neither unreasonably distant from nor unreasonably close to the date on which it was necessary for him to receive the same'.

This wording is very unfortunate, but it means that if the contractor does not apply at the right time he loses any right to an extension of time. The date of the contractor's written application must have regard to the date for completion, and we suggest that this is to be calculated in relation to the actual completion date and not to any earlier date by which the contractor hoped to complete.

Factors to be borne in mind include the state the works have reached and whether the contractor can act on the information; the nature of the instruction or information; any other of the contractor's activities which may depend on the supply of information (eg, pre-ordering of materials); and the time it may reasonably take for you to prepare the information.

5.2.3 Obedience to architect's instructions

As explained in Chapter 4, your authority to issue instructions is limited to those instructions which the contract empowers you to issue. Under clause 3.5.1, the contractor has an obligation to obey all written instructions given by you which are allowed under the contract (see Fig 4.3). He can require you to state in writing the contract clause under which the instruction is given (clause 3.5.2). He also has a right to object to instructions falling within clause 3.6.2, which is concerned with alterations or obligations or restrictions imposed by the employer in the contract documents.

The contractor must make 'reasonable objection in writing', but no guidance is given as to what is a reasonable objection. It is clearly not a reasonable objection that it is difficult for the contractor to comply, because he is expected to overcome those difficulties, and questions of time and cost are dealt with under the contract. In fact, it seems that the instruction must make continued execution of the work almost impossible: eg, by preventing deliveries to site by further restricting access.

The general sanction for non-compliance by the contractor is set out in the second part of clause 3.5.1. Under that provision, where you have served the contractor with a written notice requiring compliance and the contractor has not complied within seven days, the employer may engage others to carry out the work and deduct the cost from monies due to the contractor.

It is essential that, if the contractor does not comply with an instruction, you ensure that the remedy provided by the clause is put into operation. If you do not, there is case law that suggests that the employer may then be taken to have waived his rights under the clause

5.2 Other obligations

and, of course, your notice of compliance may be regarded as becoming stale.

More effective sanctions are provided by clause 3.13.1 (instructions following failure of work, etc) which is a most useful provision (see Fig 4.5). It covers failure of work, materials, or goods discovered during the carrying out of the works. If the contractor discovers a failure of work or of materials or goods while the works are being carried out, he must notify you in writing immediately. 'Upon such discovery [the contractor] shall state in writing to the architect the action which the contractor will *immediately* take at no cost to the employer to establish that there is no similar failure in work already executed or materials or goods already supplied', is what the clause says. This obligation extends to failures through the fault of any subcontractor, named or otherwise. The clause goes on to deal with your powers. It empowers you to issue instructions requiring opening up of work, etc, in three cases:

— where you have not received the contractor's written statement within seven days of discovery of the failure; or

— if you are dissatisfied with the contractor's proposed action; or

— if you are unable to wait for the contractor's written proposals because of considerations of safety or statutory obligation (eg, service of a dangerous structure notice or a prohibition notice under the Health and Safety at Work, etc, Act 1974).

Your default powers are to issue the appropriate instructions in writing, requiring the contractor, at his own cost, 'to open up for inspection any work covered up or to arrange for or carry out any test of any materials or goods ... or any executed work to establish that there is no similar failure and to make good in consequence thereof'. The contractor is bound to comply *forthwith* with your instructions and, while clause 3.13.2 gives him the right to object, this is said to be 'without prejudice to his obligation' to comply. In other words, even if the contractor disagrees with you, he must carry out your instruction.

Within 10 days of *receipt* of your clause 3.13.1 instructions, the contractor may object to compliance, stating his reasons in writing. You must then consider the objections immediately and decide whether to withdraw or modify the instruction to meet the contractor's objection. If you do not do so within seven days of receipt of the contractor's objection (and reasons), 'then any dispute or difference as to whether the nature or extent of the opening up for inspection or testing instructed ... was reasonable in all the circumstances' is referred to immediate arbitration. The arbitrator is given wide powers to deal with questions of both time and cost (see also section 4.1.4).

The contractor's obligations

5.2.4 Other rights and obligations

Table 5.1 summarises the contractor's powers and duties generally. Other matters referred to in that table are dealt with in the appropriate chapters.

5.3 Summary

The contractor must:
— carry out and complete the works in accordance with the contract documents
— proceed regularly and diligently with the works so as to complete in due time
— use workmanship and materials of an adequate standard
— comply with relevant statutory obligations
— obey architect's instructions as authorised by the contract
— appoint and keep on site a competent person-in-charge.

6 The employer's powers, duties, and rights

6.1 Express and implied powers and duties

Like those of the contractor, some of the employer's powers and duties arise from the express provisions of IFC84 itself. These are set out in Table 6.1.

Others are imposed by the general law by way of implied terms. These are provisions which the law writes into every building contract and apply so far as they are not excluded or modified by the express terms of the contract itself. In practice, there are two important implied terms which are not affected by the contractual provisions.

6.1.1 Co-operation or non-interference

Under the general law, it is an implied term in every building contract that the employer will do all that is reasonably necessary on his part to bring about completion of the contract. Conversely, it is implied that the employer will not so act as to prevent the contractor from completing in the time and in the manner envisaged by the agreement. Breach of either of these implied terms which results in loss to the contractor will give rise to a claim for damages at common law.

Equally, if the employer—either personally or through your agency or that of anyone else for whom he is responsible in law—hinders or prevents the contractor from completing in due time, not only is he in breach of contract, but conduct of this sort will also prevent him from enforcing the liquidated-damages provision if any delay results.

The various cases put the duty in different ways, but in essence the position may be summarised as follows (*at the head of page 82*).

The employer's powers, duties, and rights

Table 6.1 Employer's power and duties under IFC84

Clause	Power/duty	Preconditions/comment
1.6	**Duty** Be custodian of the contract documents	Contract documents must be available for contractor's inspection at all reasonable times
1.8	**Duty** Not to divulge or use any of the contractor's rates and prices	Except for the purposes of the contract
2.1	**Duty** Give possession of the site to the contractor on the date for possession **Power** Defer giving possession of the site to the contractor for a limited period	Where clause 2.2 is stated in the appendix to apply Deferment is for a period not to exceed stated period; usual maximum period is six weeks
2.7	**Power** Deduct liquidated damages for late completion	Architect must have issued a certificate of non-completion *and* employer must have required liquidated damages by writing to the contractor not later than the date of the final certificate
2.8	**Duty** Pay or repay liquidated damages to the contractor	Where the architect cancels his certificate of delay and grants a further extension of time after issuing it
2.10	**Power** Consent to defects not being made good	An appropriate deduction is to be made from the contract sum
3.1	**Power** Assign the contract	Only if the contractor consents in writing
3.3.1 & 3.3.4	**Power** Have a named person's work carried out by his employees or direct contractors under clause 3.11	Where contractor has been unable to subcontract with the named person *and* the architect has instructed omission of the work or has so omitted it and substituted a provisional sum. Such instructions are valued as a variation and give rise to a contractor's claim for both extension of time and loss and/or expense
3.3.6	**Power** Indemnify the contractor against legal costs	If the employer requires the contractor to commence legal or arbitral proceedings against a defaulting named person

78

6.1 Express and implied powers and duties

3.5	**Power** Employ and pay other persons to execute work	If the contractor does not comply within seven days of the receipt of a written notice from the architect requiring compliance with an instruction
3.7	**Power** Agree with the contractor the amount of an adjustment to the contract sum	In respect of variation or provisional sum instructions, and before the contractor complies therewith
3.9	**Power** Consent to the architect's instructing the contractor that setting out errors be not amended	An appropriate deduction is to be made to the contract sum
3.10	**Power** Appoint a clerk of works to act as an inspector under the directions of the architect	
3.11	**Power** Require work not forming part of the contract to be carried out by himself or by persons employed or engaged by him	Where the contract documents so provide, *or* with the contractor's consent
3.13	**Duty** Pay the contractor any amount awarded by the arbitrator in respect of compliance with architect's instructions following failure of work, etc	If the contractor has objected in writing to compliance, with reasons, and the architect has not withdrawn or modified his instruction, and the matter has been pursued to arbitration
4.2	**Duty** Pay to the contractor the amount certified within 14 days of the date of the certificate	If the architect issues an interim certificate under clause 4.2
4.3	**Duty** Pay to the contractor the amount certified within 14 days of the date of the certificate	If the architect issues an interim certificate under clause 4.3. On practical completion, a further interim payment to bring the amount to 97½ per cent is made
4.4	**Power** To have recourse to the percentage retained from time to time for payment of any amount to which he is entitled under the contract provisions to deduct from sums due, or to become due, to the contractor	Where the employer is not a local authority
4.6	**Duty** Pay to the contractor the amount certified within 21 days after the date of the final certificate	Subject to any amounts properly deductible
5.5	**Duty** Pay to the contractor any VAT properly chargeable	

79

The employer's powers, duties, and rights

Table 6.1
Employer's power and duties under IFC84—*cont.*

Clause	Power/duty	Precondition/comment
6.2	**Power** Insure against property damage caused by collapse, subsidence, etc	Where a provisional sum is included in the contract documents and the contractor has made default in insuring or in continuing to insure
6.3A	**Duty** Pay insurance monies to the contractor **Power** Insure against clause 6.3 perils	Where certified by the architect under clause 4.2. The payment is net of any percentages to cover professional fees Where the contractor has made default in insuring or containing to insure
6.3B	**Duty** Maintain proper insurance against the clause 6.3 perils	Where the employer has undertaken the risk in the case of new works. (A local authority employer is not under this duty.)
6.3B	**Duty** Produce insurance receipts	On being so requested by the contractor
6.3C	**Duty** Maintain adequate insurances against clause 6.3 perils **Duty** Produce insurance receipts **Power** Determine the contractor's employment under the contract	In the case of existing structures If contractor so requests If the works are damaged by clause 6.3 perils and if it is just and equitable to do so
7.1	**Power** Serve written notice on the contractor by registered post or recorded delivery, specifying a default **Power** Determine the contractor's employment by written notice served by registered post or recorded delivery	If the contractor without reasonable cause wholly suspends the carrying out of the works before completion *or* fails to proceed regularly and diligently with the works; *or* refuses or persistently neglects to comply with a written notice from the architect requiring him to remove defective work or improper materials or goods and by such refusal or neglect the works are materially affected; *or* fails to comply with clauses 3.2 (subcontracting), 3.3 (named persons), or 5.7 (fair wages) The notice must not be given unreasonably or vexatiously. It can be served only if the contractor continues his default for 14 days after receipt of the preliminary notice or if at any time thereafter he repeats that default

6.1 Express and implied powers and duties

7.2	**Power** Reinstate the contractor's employment in agreement with the contract and his trustees (in bankruptcy), liquidator, etc	Where the contractor has become insolvent, etc; his employment is determined automatically
7.3	**Power** Determine the contractor's employment under this or any other contract	Where the employer is a local authority and the contractor is guilty of corrupt practices. No procedure is prescribed, but determination should be effected by written notice
7.4	**Duty** Pay to the contractor any amount due to him after completion of the works by others	
7.7	**Duty** Pay to the contractor the total value of work at the date of determination; sums ascertained as direct loss and/or expense under clause 4.10; the cost of materials or goods properly ordered for the works for which the contractor has paid or is legally bound to pay; the reasonable cost of removal from site of all temporary buildings, plant, tools, equipment, etc; and any direct loss and/or damage caused to the contractor by the determination	Where the contractor has determined his own employment for the employer's default or insolvency under clauses 7.5 or 7.6 Amounts previously paid are taken into account Direct loss and/or damage will include the contractor's loss of profit
7.8	**Power** Forthwith determine the contractor's employment by written notice served by registered post or recorded delivery	Where the carrying out of the whole or substantially the whole of the uncompleted works is suspended for three months by reason of *force majeure*, or loss or damage caused by clause 6.3 perils (eg, fire), *or* civil commotion The notice must not be given unreasonably or vexatiously
7.9	**Duty** Pay to the contractor the total value of work at the date of determination; any sum ascertained as direct loss and/or expense under clause 4.10; the cost of materials or goods properly ordered for the works for which the contractor has paid or is legally bound to pay; the reasonable cost of removal from site of all temporary buildings, plant, tools, equipment, etc	Amounts previously paid are taken into account

— the employer and his agents must do all things necessary to enable the contractor to carry out and complete the works expeditiously in accordance with the contract.

— neither the employer nor his agents will in any way hinder or prevent the contractor from carrying out and completing the works expeditiously and in accordance with the contract.

The scope of these implied obligations is very broad, and in recent years more and more claims for breach of them have been before arbitrators or the courts. The employer must not, for example, attempt to dictate to you how to exercise your discretion, nor must he attempt to give direct orders to the contractor. Similarly, he must see that the site is available for the contractor and that access to it is unimpeded by those for whom he is responsible. This is especially important in works to existing structures or tenanted buildings, and these are matters which you should discuss with the employer before the contract is let.

Some potential acts of hindrance or prevention by the employer are covered by express clauses in the contract, but there are a number of grey areas.

6.2 Rights

6.2.1 General

Although the contract is between the employer and the contractor—who are the only parties to it—an analysis of the contract clauses shows that the employer has few express rights of any substance.

The employer's major right is, of course, to have the completed works handed over to him in due time, properly completed in accordance with the contract documents. But his other rights are of importance as the contract proceeds.

6.2.2 Deferment of possession of the site

Clause 2.2 confers on the employer a right which he would not otherwise possess, namely the right to defer giving possession of the site to the contractor for a period up to six weeks (assuming that the appendix states that this provision is to apply). This is an important right in practice because, under the general law, failure to give the contractor sufficient possession to enable him to proceed with the works is a serious breach of contract.

This power will be especially helpful in renovation works, for example, although the normal intention must be that it is to be exercised sparingly, since if one has reached the contract stage, it is to be

6.2 Rights

assumed that sufficient possession will be given to the contractor on the due date. Certainly, in the absence of such a provision, there would be no power to defer or postpone giving possession, and it would be necessary for the employer and the contractor to reach a separate agreement.

6.2.3 Deduction/repayment of liquidated damages

If the contractor is late in completing the works, then—provided you have issued your certificate of non-completion under clause 2.6—the employer is entitled to recover liquidated damages at the rate specified in the appendix. This is usually done by deduction from sums due to the contractor (eg, under interim certificates), but if (unusually) no sums are due to the contractor, then the employer must sue for them as a debt.

There is a further precondition to deduction or recovery. Clause 2.7 makes it plain that payment of liquidated damages is not obligatory. The mere fact of late completion and the issue of your certificate is not sufficient. The employer must give notice to the contractor, before the issue of the final certificate, of his intention to exercise his discretion to claim or deduct liquidated damages, because he is required to indicate to the contractor whether he wants them allowed or to be paid.

You should advise the employer of his rights, and draft a suitable letter for him to send to the contractor. Fig 6.1 is a suitable pro forma letter. It is a moot point whether the employer can recover ordinary unliquidated damages for late completion if he has not exercised his right to deduct or has failed to give the requisite notice to the contractor. On general principle, the answer seems to be that he can recover unliquidated damages in these circumstances by way of legal action or in arbitration, and these may amount to much more than the fixed (liquidated) sum. The corresponding disadvantage is that, in such a case, the employer will need to prove his actual loss.

The employer must repay any liquidated damages which he has recovered should you cancel your clause 2.6 certificate and grant a further extension of time. Undoubtedly, contractors will argue that any damages so repaid should attract interest, and many ingenious arguments are likely to be advanced to support this contention, notably with reference to a decision of the High Court of Northern Ireland. Specialist practitioners are generally agreed that, even if correct on its facts, that decision is not good law generally—and it certainly has no application to IFC84. If interest was payable, this would be stated in clause 2.7. Should the contractor advance such an argument. Fig 6.2 is the sort of reply which you would advise the employer to send. You have nothing to do with the deduction or enforcement of liquidated damages.

The employer's powers, duties, and rights

Fig 6.1
Employer to contractor, regarding deduction of liquidated damages

REGISTERED POST OR RECORDED DELIVERY

Dear Sir

The architect having issued his certificate of non-completion under clause 2.6 of the above contract, certifying that you have failed to complete the works by [*insert date as certified*], in accordance with clause 2.7 of the contract, I hereby require that you pay or allow liquidated damages at the rate of £ [*insert amount*] for every week or part of a week during which the works remain uncompleted/have remained uncompleted [*omit as appropriate*].

Yours faithfully

Fig 6.2
Employer to contractor if contractor claims interest on liquidated damages repaid

Dear Sir

Thank you for your letter of [*insert date*] (addressed to the architect, Mr/Messrs [*insert name*]) in which you claim that you are entitled to interest on the liquidated damages repaid/which will be repaid [*omit as appropriate*] to you as a result of the architect's cancelling the current/original [*omit as appropriate*] clause 2.6 certificate.

I/we formally deny that you are entitled to the interest claimed or to any interest at all. The contract makes no provision for the payment of interest in these circumstances, and interest is not recoverable under the general law.

Yours faithfully

6.2.4 Employment of direct contractors

Clause 3.11 is an important provision, since under it the employer has the right—if the contract documents so provide—to carry out himself or have carried out by others 'work not forming part of this contract'. The contractor is obliged to permit this work to be executed while the contract works are in progress, provided that reference is made to it in the contract documents. If no such reference is made, the employer can still have such work carried out with the consent of the contractor, which must not be unreasonably withheld.

This is an important right which the employer certainly would not have at common law, since, as explained in Chapter 9, the contractor is in principle entitled to exclusive possession of the site during the currency of the contract. But the limitations of the provision must be noted.

— it applies only to work 'not forming part of this contract' and cannot be used to take away from the contractor work which is his.

This is subject to the limited provision in clauses 3.3.1 and 3.3.4 in relation to the work and named subcontractors where the contractor has, for good reason, been unable to enter into a subcontract with the person named and you have either instructed its omission or have omitted the work from the contract documents and substituted a provisional sum. In either event, the contractor is entitled to extension of both time and money.

The fundamental principle is that a clause of this sort cannot in general be used to omit contract work to give to others, because the contractor has agreed to do a certain quantity of work and has a right to do it. But a provision in terms of clause 3.11—sometimes called an 'Epstein clause'—does enable the employer to carry out work himself while the contract is in progress—eg, through his own direct works department—or to have similar work carried out by others.

The words 'not forming part of this contract' mean exactly what they say. It is not work which the employer can require the contractor to do, and it is important to get this matter clear because of the implications in time and cost. Under clauses 2.4.8 and 4.12.2, the contractor is entitled to both extension of time for any consequent delay and also to a loss and/or expense claim. Statutory undertakers—water boards, gas boards, and the like—may fall under this clause if they are in a direct contract with the employer, but not where they are carrying out work 'in pursuance of [their] statutory obligations', and this is a point about which you should be most careful.

It is best to discuss the employer's requirements with him at an early stage and explain the situation to him in terms of both time and cost. You would be well advised to write an appropriate provision into the contract documents rather than rely on the contractor's consenting subsequently.

The employer's power directly to employ and pay others to execute the work (clause 3.5) where the contractor has failed to comply with your notice requiring compliance with an instruction does not fall under clause 3.11.

The use of clause 3.11 may be one method of avoiding some of the difficulties thrown up by clause 3.3 with regard to named subcontractors, since by careful forethought, specialist work can be made the subject of direct employment by the employer; in that case, the employer is solely responsible for the acts or defaults of such subcontrators and has no action against the contractor, whether for insurance purposes or otherwise.

6.2.5 Rights as to insurance

The complex insurance clauses of IFC84 were discussed in Chapter 3, but the employer's rights relating to insurance should be noted.

Clause 6.2.3 gives the employer a default power to insure or maintain policies in force where the contractor (or any subcontractor of his) fails to do so and to recoup himself out of the monies due or to become due to the contractor. A provisional sum for the insurance must have been included in the contract documents if he is to rely on this provision.

Clause 6.3C—Special power of determination (see section 12.1.6): the power to determine the contractor's employment can be exercised only if it is just and equitable so to do—a matter which the arbitrator may have some difficulty in deciding.

The employer's other contractual rights are summarised in Table 6.1, where they are described as 'powers'. They are discussed in the appropriate chapters.

6.3 Duties

6.3.1 General

The essence of a duty is that it must be carried out. It is not permissive but mandatory, and breach of a duty imposed by the contract will render the employer liable in damages to the contractor for any proven loss.

Not every breach of a contractual duty will entitle the contractor to treat the contract as being at an end: only breach of a provision which goes to the root or basis of the contract will do that. But a breach of any contractual duty will always, in theory, entitle the contractor to at least nominal damages, although in many cases any loss will be difficult if not impossible to quantify.

6.3.2 Payment

From the contractor's point of view, the most fundamental duty of the employer is to make payment in accordance with the terms of the contract. However, while steady payment of certificates is an essential from the contractor's point of view, the general law does not regard failure to pay, or to pay on time, as a major breach of contract.

IFC84 is quite specific about payment (see Chapter 11). The basic provisions are contained in clauses 4.2 and 4.3 (interim payments) and clause 4.6 (final payment), although the contractor's remedies for non-payment of an interim certificate are limited. He can, of course, sue on the certificate once payment is due, but he is not entitled to interest on the overdue sum. He certainly has no right to suspend work, although, in an extreme case and after going through the specified procedure, he has the right to determine his employment under clause 7.5.1 as there set out (see Chapter 12).

Once you have issued your interim certificate (clauses 4.2 and 4.3), the employer is given a period of grace before he need honour the certificate. He must make payment within 14 days of the date of the certificate, which means payment before the expiry of that period. If you send your certificate to the employer by post, then effectively he has 13 days for paying the amount due—assuming, of course, that the certificate is delivered the next day.

Payment by cheque is probably good payment, although some contractors have been known to argue to the contrary under other forms of contract. It is not permitted for the employer to say, for example, that his computer arrangements do not fit in with the scheme of certificates. If this is so, then the payment period should have been amended before the contract was let. You should emphasise to the employer the need to honour certificates promptly.

For the final payment (clause 4.6), the period of grace is 21 days, since the last sentence refers to the certified amount being a debt payable 'as from the twenty-first day after the date of the final certificate'. By the terms of the contract, the employer is entitled to deduct from both interim and final certificates certain specific and ascertained sums—notably, liquidated and ascertained damages under clause 4.7.

The contract also contains other express provisions empowering deductions. These are clause 3.5.1—costs incurred in employing others to give effect to an architect's instruction; clause 6.2.3—insurance premiums paid on contractor's default; clause 6.3A.2—insurance premiums similarly paid; clause 7.4—payments on determination of employment; supplemental condition B—VAT payments as specified therein.

6.3 Duties

6.3.3 Retention

The employer has certain rights in the retention percentage. The retention monies are a trust fund except where the employer is a local authority, as is manifest by the wording of clause 4.4 which reads as follows. 'Where the employer is not a local authority the employer's interest in the [retention percentage] shall be fiduciary as trustee for the contractor (but without obligation to invest) and the contractor's beneficial interest therein shall be subject only to the right of the employer to have recourse thereto from time to time for payment of any amount which he is entitled under the provisions of this contract to deduct from any sum due or to become due to the contractor.'

The deductions which the employer may make from the percentage retained are those referred to in section 6.3.2. Since the retention money is trust money, it is not the employer's property—hence the need to confer on him the contractual right to deduct from it. This apart, the employer has no legal or other interest in the retention and, it seems, the court can order him—on the contractor's application—to set the percentage withheld aside in a separate bank account, although this would be unusual. Nothing is said about whether the contractor is entitled to interest on the percentage withheld, and it has been convincingly argued about a similarly worded clause that he is entitled to interest since the employer is in the position of a trustee of the money.

6.3.4 Other duties

These are summarised in Table 6.1 and are commented on as necessary in appropriate chapters.

6.4 Summary

The employer is under a duty in common law to do all that is reasonably necessary to bring about completion of the contract. Under the contract, the employer has a right to:
— defer giving possession of the site to the contractor for a maximum period of 6 weeks
— recover liquidated damages for late completion
— employ direct contractors to carry out works 'not forming part of the contract'
— insure in default of the contractor so doing.

The employer must as a duty:
— pay the contractor in accordance with the contract terms
— observe all the contract provisions.

7 The clerk of works

7.1 Appointment

The appointment of a clerk of works is a matter for the employer acting on your advice. Some organisations have their own clerks of works on their permanent staff. Some firms of architects employ clerks of works to provide a service of frequent or constant inspection, but in the light of recent case law (see section 7.3), it is not advisable for you to employ the clerk of works. It is much better, from your point of view, if he is employed directly by the employer.

Whether a clerk of works is necessary on a full-time or part-time basis or at all is a matter about which you must decide and advise the employer. If you have entered into a form of agreement with the employer incorporating the *Architect's appointment*, clause 3.10 of that document makes it clear that you will not be required to make frequent or constant inspections. If such inspections are required, clause 3.11 states that a clerk of works will be employed.

You should ask the employer to appoint the clerk of works as soon as possible. That will be immediately after he has accepted a tender for the work. The clerk of works then has time to become thoroughly acquainted with the work which is to be done. Make sure that you give the clerk of works a letter outlining what you require him to do. You will, of course, brief him thoroughly in your office, but a letter serves to remind him of the main points you consider to be important. Every job is different and every architect has different ideas about his relationship with the clerk of works, but among the matters you might include in your letter are the following:

— his general duty to inspect

7.2 Duties

— how you wish him to deal with defects he discovers
— that he has no authority to issue instructions
— how you wish him to deal with queries
— filling in of report sheets
— filling in of diary
— any purpose for which you intend to make him your authorised representative
— confirmation that he has received copies of all the documents issued to the contractor, together with diary, report forms, stationery, etc.

7.2 Duties

The duties of the clerk of works are set out in the contract, clause 3.10: 'The Employer shall be entitled to appoint a clerk of works whose duty shall be to act solely as an inspector on behalf of the Employer under the directions of the Architect'. The contract makes no other reference to him. He is not given any power to issue even directions (as in JCT80).

His only function is to inspect. The clause makes it clear that he carries out this function on behalf of the employer, not on your behalf, but you and not the employer are empowered to direct him. Since he may only inspect, the subject of your directions must, presumably, be how and where he should inspect, at what intervals, and to what he should pay particular attention. Although it is traditional to direct the clerk of works in a fairly informal manner, you should always give him written confirmation of any directions which are other than routine. This will help to protect your position later. It is generally accepted, by architects, contractors, and clerks of works alike, that the clerk of works will usually do more than simply inspect. You will often ask him to take measurements or levels for you, and the contractor may rely on him to solve problems on site. It must be understood, however, that the contract recognises none of this. If the clerk of works issues any instructions to the contractor, the contractor would be well advised to ignore them. He cannot successfully claim reimbursement for carrying out any instructions of the clerk of works, no matter how sensible they appear to be. Of course, the clerk of works may make quite useful suggestions to you. He will often give you a valuable service by finding mistakes on your drawings, suggesting ways of saving money, and generally assisting in ensuring the smooth running of the contract on site. Make sure, at the very beginning of the contract, that the contractor understands precisely the role of the clerk of works. It is a good idea to do this at the first meeting when the clerk of works is present and record it clearly in the minutes. If you do not clarify things in this way, do not expect the contractor to have a copy of clause 3.10

The clerk of works

pinned up on site; he will not necessarily realise how strictly you intend to apply the contract provisions. It will be a disaster if he finds out because you overrule the clerk of works in some way.

Many commentators, including the Institute of Clerks of Works, consider the clerk-of-works provisions in JCT80 inadequate. They will probably take a similar view of IFC84. You can, of course, advise the employer to amend these provisions at tender stage if you wish, but get expert legal advice regarding the possible consequences. It seems that they should be left exactly as they stand or be made much more comprehensive.

The easiest way for the clerk of works to view his duties is to split them in two:

— his duties under the contract to inspect only;
— his duties by virtue of your directions to him.

In general terms, his duties under the contract define his relations with the contractor, while his duty to comply with your directions affects only his relationship with you. If you can get the clerk of works to see his job in this light, you should remove most problems.

It is common practice for the clerk of works to make marks in chalk or wax crayon to indicate defects. It has been known for a clerk of works to deliberately deface unsatisfactory materials and goods to ensure that they are not incorporated or, if already incorporated, to ensure that they are replaced. The clerk of works is not entitled to take this kind of action. If he does, the employer could find himself facing a large bill from the contractor. The clerk of works would probably argue that he cannot possibly do any harm to something which is already defective. He should remember that defective materials are not, by definition, the property of the employer and that the contractor may, and usually does, find another use for them elsewhere.

It is also bad practice for the clerk of works to get into the habit of putting specific marks on work or materials for the same reason that he should not issue what are usually termed 'snagging lists'. The clerk of works is on site for the benefit of the employer. Naturally, he must point out major defects. Similarly, he must draw the contractor's attention to the host of minor defects which are often present. But being more specific has two dangers. One is that the contractor may consider that, if he rectifies everything on the clerk of works' list, he has fulfilled his obligations, and that dispute, however misguided, may follow. The second is that the clerk of works is doing the job which properly belongs to the person in charge.

7.3 Responsibility

The clerk of works has a responsibility to carry out his duties in a competent manner. He is expected to show the same degree of skill as would be shown by the average clerk of works. If he holds himself out as being specially qualified in some branch of the industry, greater skill will be expected of him in that respect.

In spite of the reliance placed on the clerk of works by employer and architect, it was not clear, until recently, whether and to what extent he was accountable for his actions. Recent case law has settled the matter, for the moment. Provided that the clerk of works is employed to inspect on behalf of, and is paid by, the employer, the employer is vicariously liable for his actions. This is so although he is under your directions. The ordinary relationship of master and servant must exist. Therefore, if *you* employ the clerk of works or he is permanently on your staff, the position will remain uncertain. This does not mean to say that, if the clerk of works is negligent, the fact that he is the employer's servant will relieve you of all responsibility, though it may reduce your liability for damages, depending on the circumstances. It is, therefore, very much in your interest to ensure that the employer employs a clerk of works.

7.4 Summary

— the clerk of works should be appointed by the employer
— he will be necessary if frequent or constant inspections are required
— he should be appointed immediately the successful tender has been accepted
— he should be thoroughly briefed by you
— he is purely an inspector
— he may carry out other duties for you
— he is not entitled to put any mark on the works
— he should be discouraged from producing 'snagging lists'
— he must carry out his duties competently
— he may reduce your liability for damages if the employer is found to be vicariously liable.

8 Subcontractors and suppliers

8.1 General

This chapter deals with the contract provisions for subcontractors, suppliers, statutory authorities, and persons engaged by the employer to carry out work not forming part of the contract.

There are no provisions in IFC84 for nominating subcontractors or suppliers. Provision is made for the employer to name persons to carry out work priced in the contract documents or included as a provisional sum (see section 8.2). There are no similar provisions for suppliers.

8.2 Subcontractors

8.2.1 Assignment and subcontracting

IFC84 contains the usual restriction (clause 3.1) on the assignment of the contract by either party without the written consent of the other. Without this express term, it would be possible for either party to assign the *benefits* of the contract to another. For example, the contractor might wish to assign to a third party the benefit of receiving payments under interim certificates, in return for which the third party would give the contractor financial advances to enable him to carry out the work; this is a well known procedure. Or the employer might wish to sell the building before the issue of the final certificate or even when the structure was only partly completed. Even without the express term, it is a matter of general law that the *burden* of a contract cannot be assigned without the consent of the other party. For example, the

8.2 Subcontractors

contractor cannot transfer to another the burden of carrying out the work and the employer cannot pass to another his obligation to pay the contract sum.

There are very real difficulties for both parties in this clause, since the party withholding consent is not required to be reasonable in doing so. It is your duty to warn the employer and seek his instructions on amending the clause if he is likely to want to assign his benefit before the time for issuing the final certificate.

In contrast, if the contractor wishes to sublet any part of the works, you must not withold your consent unreasonably. The contractor does not have to inform you of the name of the subcontractor, but you would probably be reasonable in withholding your consent if he did not do so (clause 3.2). There is a proviso (clause 3.2.1) that the subcontract must provide for the employment of any subcontractor to determine immediately on the determination of the contractor's employment for whatever reason. This is a perfectly sensible provision to prevent the situation arising in which the contractor's employment is determined and the subcontractor is able to sue the contractor because he is prevented from carrying out the subcontract. It must be remembered that the provisions of a contract cannot bind anyone who is not a party to it. Therefore, this clause places the onus on the contractor to ensure that an appropriate term is included in the subcontract.

The employer has no contractual relationship with the subcontractor, and the contractor must bear liability for any defects in the subcontractor's work. The employer will look to the contractor for redress in such circumstances. The contractor must, in turn, look to the subcontractor. Any dispute, difficulty, or difference arising between contractor and subcontractor is a matter solely for the parties involved. You should beware of being drawn into such disputes.

Clause 3.2.2 sets out certain provisions which the contractor must ensure are included in any subcontract. It would not be unreasonable if you made your consent to subletting subject to the contractor's providing documentary evidence that these provisions are to be so included. They relate to the ownership of unfixed materials on site. The intention is to prevent the subcontractor from reclaiming goods delivered to site which have already been paid for by the employer under your certificate. Subsection (a) of the clause requires the contractor's consent to the subcontractor's removing such materials from site, and this subsection is made subject to main contract clause 1.10 which requires your consent (see section 4.2.1). Effectively, therefore, the subcontractor requires your consent, through the contractor, before unfixed materials can be moved. The consents must not be withheld unreasonably.

Subsection (b) provides that the materials will become the property of

Subcontractors and suppliers

the employer if he pays the contractor under a certificate which includes the value of such materials. The subcontractor is not to deny that the materials are the employer's property. Note that this provision takes no account of the fact that the contractor may not have paid the subcontractor. If the contractor does pay the subcontractor before he has been paid himself, subsection (c) provides that the materials become the property of the contractor. Subsection (d) makes it clear that the other parts of this clause are not to have the effect of preventing the employer from acquiring ownership of materials in accordance with clause 1.11.

A similar set of provisions is now included in JCT80. They were prompted by a recent case in which the employer paid for subcontract materials, but, before the contractor paid the subcontractor, the contractor went into liquidation. The goods were held to be the property of the subcontractor, and the employer had to pay for them again. Whether the current provisions are effective in preventing a recurrence of this problem, time will tell. It may be expected that subcontractors will increase their prices to cover the risk that they now appear to take.

However, it must be emphasised that these provisions will not necessarily protect the employer. For example, if the subcontractor's materials have been supplied to him on terms of sale including a retention-of-title clause, this would not be defeated by clause 3.2.2 or any corresponding provision in the subcontract.

8.2.2 Named persons as subcontractors

These provisions (clause 3.3) are new. They are also quite complicated and, in places, unclear. It is important to remember that they are not the same as the provisions for nominating subcontractors under JCT80.

A named person my be involved in one of two ways. One is if work is included in the contract documents to be priced by the contractor and carried out by a named person. In this case, the person is named in the contract documents, and the contractor does not have a choice (eg, one of three). The second is if work is included in an architect's instruction regarding the expenditure of a provisional sum and a person is named to carry it out.

The consequences are slightly different. The flowchart, Fig 8.1, outlines the procedures. In the first instance, the contractor has 21 days from entering into the main contract to enter into a subcontract with the named person. That is 21 days from the date of the employer's acceptance of his tender, not 21 days from signing the contract documents. It is a very short period. The subcontract must consist of section III of the form of tender and agreement NAM/T (This

8.2 Subcontractors

incorporates subcontract conditions NAM/SC). Section I of the form should have been completed by you, section II by the named person. You will have to have these documents ready to hand to the contractor on acceptance of his tender, or he can, quite rightly, say that you have delayed him. Moreover, this eventuality hardly comes within clause 2.4.7, which refers to 'instructions, drawings, details, or levels', so you appear to have no contractual power to award an extension of the contract period. Since you must protect the employer's interest by awarding an extension, you will be obliged to ask the employer to defer possession under clause 2.2 and award an extension under clause 2.4.14. Unfortunately, the contractor can then claim loss and/or expense under clause 4.11(a).

The clause provides that if the contractor is 'unable to enter into a subcontract in accordance with the particulars given in the contract documents', he must immediately inform you, specifying which particulars have caused the problem. What is meant by 'particulars' is nowhere defined, but they are referred to in the first and second recitals. They are to be in the contract documents and are not necessarily confined to the form of tender and agreement NAM/T, though this appears to have been the intention of the JCT. If you are satisfied that the particulars specified have indeed prevented the execution of the subcontract (and you can be sure only if the contractor sends you a letter from the subcontractor stating as much), you have three courses: alter the particulars to remove the subcontractor's objection, which creates a variation (3.3.1(a)); *or* omit the work altogether, which also creates a variation, (3.3.1(b)); *or* omit the work from the contract documents, substituting a provisional sum for which you must then issue an instruction under clause 3.3.2 (3.3.1(c)).

This clause might appear to give you the power to vary the terms of the contract, which clearly only the employer can do with the consent of the contractor. It must therefore be assumed that your power is limited under clause 3.3.1(a) to the varying of other than contract terms. If you omit the work altogether, the employer may get someone else to do the work and pay direct, subject to the provisions of clause 3.11 (see section 8.4).

The contractor must notify you of the date when he has entered into a subcontract, assuming that all goes well. Before he does so, you may issue an instruction similar to clause 3.3.1(c).

Clause 3.3.2 deals with the procedure if a named person arises through a provisional sum. This can occur in three ways: if you issue an instruction (clause 3.3.1(c)) after the particulars are said to be preventing execution of the subcontract; or if you issue a similar instruction (clause 3.3.1) omitting the work from the contract documents and substituting a provisional sum before the subcontract is

Subcontractors and suppliers

Fig 8.1
Named persons as subcontractors procedure (clause 3.3)

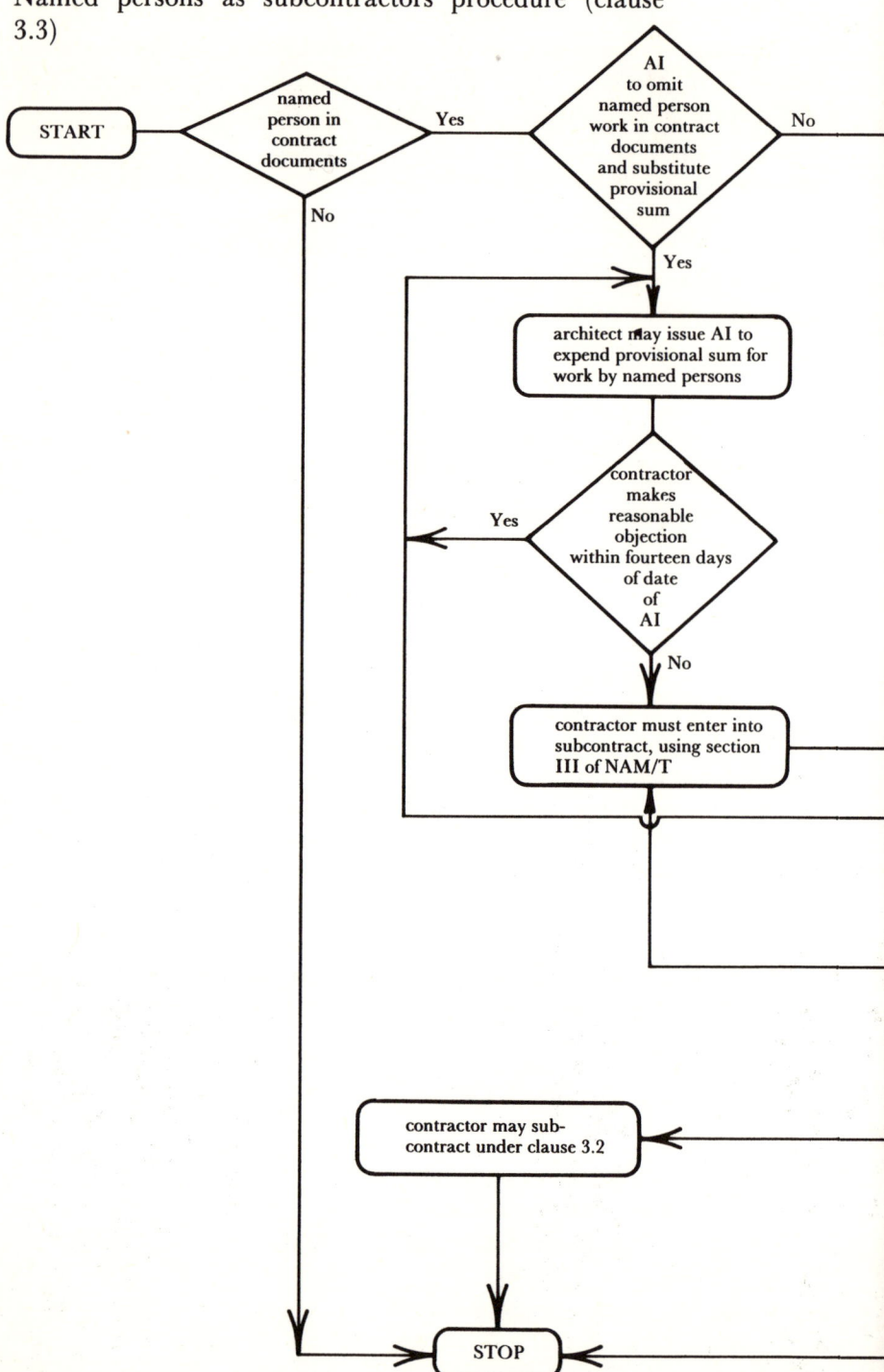

8.2 Subcontractors

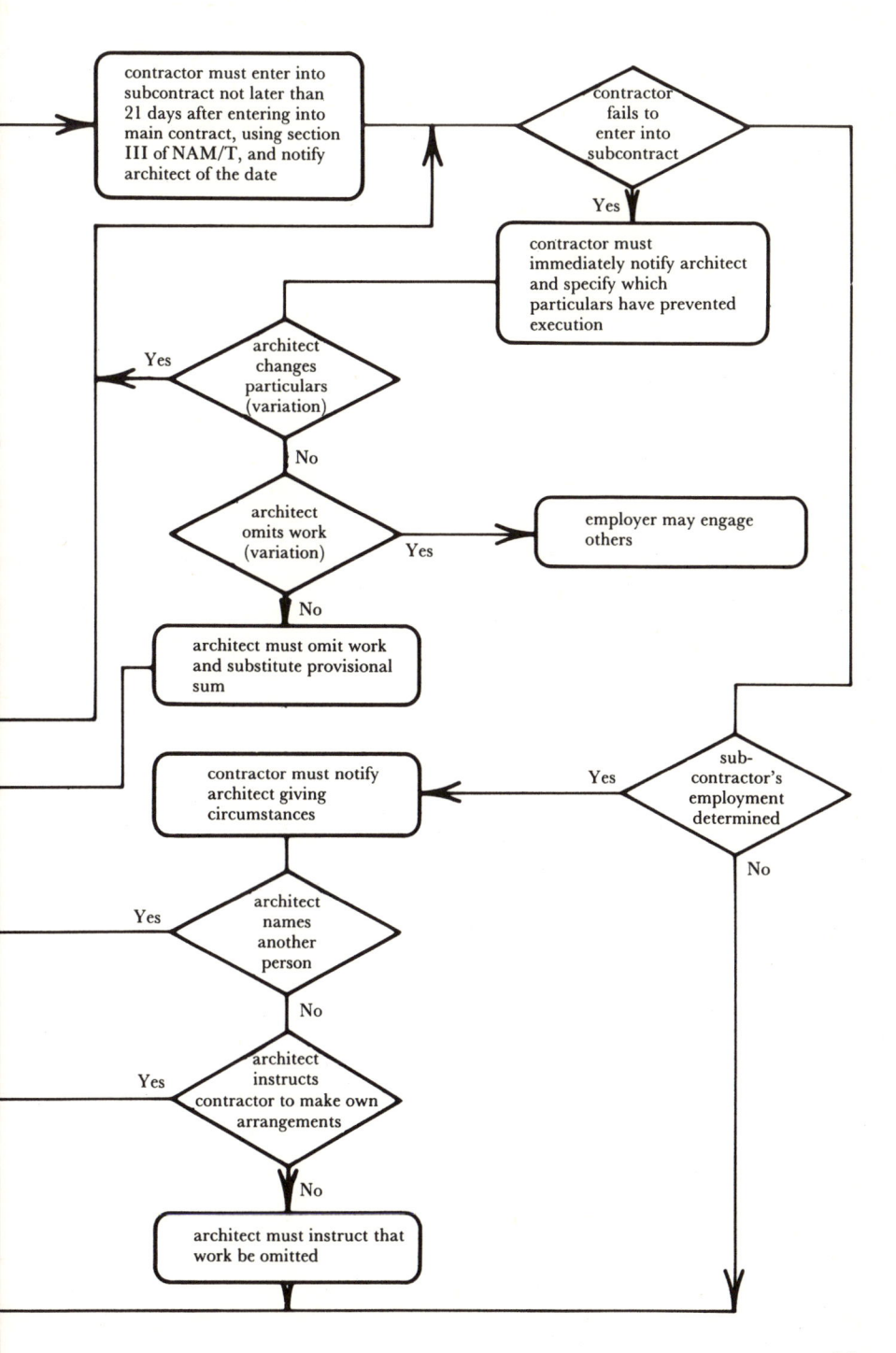

- contractor must enter into subcontract not later than 21 days after entering into main contract, using section III of NAM/T, and notify architect of the date
- contractor fails to enter into subcontract? **Yes** →
- contractor must immediately notify architect and specify which particulars have prevented execution
- architect changes particulars (variation)? **Yes** → (loop back)
- architect omits work (variation)? **Yes** → employer may engage others
- **No** → architect must omit work and substitute provisional sum
- sub-contractor's employment determined? **Yes** → contractor must notify architect giving circumstances
- architect names another person? **Yes** → (loop back)
- **No** → architect instructs contractor to make own arrangements? **Yes** → (loop back)
- **No** → architect must instruct that work be omitted

99

Subcontractors and suppliers

signed; or if there is a provisional sum in the contract documents (clause 3.8). This clause allows you to issue an instruction regarding the expenditure of a provisional sum and to require that the work is to be done by a named person. The named person is to be employed by the contractor as a subcontractor (clause 3.3.2(a)).

Your instruction must describe the work to be done and include all the details of sections I and II of the completed form of tender and agreement NAM/T for the work (clause 3.3.2(b)).

Clearly, there is a difference between a person's being named in the contract documents, when the contractor has the opportunity to see who it is before he tenders for the whole contract, and being named in an architect's instruction, when the job is in progress and the contractor is committed. For that reason, the contractor has 14 days from the date of the issue of the instruction in which to make an objection. The objection must be reasonable, the contract does not say what is to happen if the contractor does make a reasonable objection. Presumably, you must name another person and so on, if the contractor continues to object, until the contractor stops objecting. Under such circumstances, the naming of a person is not a quick process. You have to complete section I of the form of tender and agreement NAM/T and send a copy to each of the persons you wish to tender. Each tenderer must complete section II and return it to you. Assuming that you have an inexhaustible supply of suitable tenderers, the problem is the delay to the contract. You have no power to award an extension of time. Clause 2.4.5 refers to compliance with architect's instructions 'to the extent provided therein, under clause 3.3'. Clause 3.3 does not provide for an extension of time for delay caused by the contractor's making reasonable objection. You may postpone the work under clause 3.15 and grant an extension under clause 2.4.5. The contractor will then claim loss and/or expense under clause 4.12.5.

When you have got the contractor's objections sorted out, or if he does not make any objection, he must enter into a subcontract with the named person, using section III of the form of tender and agreement NAM/T. If the contractor is unable to enter into a subcontract because of the 'particulars', the contract is again silent regarding the next move. What is certain is that the employer has the responsibility of naming a person who will enter into a subcontract on the basis of the particulars. The best way to settle this situation is by negotiation. You should attempt to come to some agreement about the particulars with all parties. If that fails, name another person. In practice, these sort of problems are unlikely to occur while all the participants are equally bewildered by these provisions. Within a short time, however, the contractors and subcontractors will become familiar with the loopholes, and you could have difficulties until the provisions are

8.2 Subcontractors

revised. The employer can always arrange to amend this clause, as any other, at tender stage, but make sure he does so with expert legal advice.

It is essential that you advise the employer to enter into the RIBA/CASEC Form of Agreement between the Employer and a Specialist to be named as a subcontractor under the JCT Intermediate Form of Building Contract. This will enable the employer to take direct action against the named person if he defaults, in particular if he fails in any of the following:

— any design of the subcontract works which he has undertaken to do;
— the selection of the kinds of materials for the subcontract works which he has undertaken to do;
— the satisfaction of any performance specification or requirement relating to the subcontract works which he has undertaken to fulfil.

Clause 3.3.7 of IFC84 expressly removes the contractor's liability for the above items 'whether or not' the named person is responsible to the employer for them. Moreover, the named person is not to be liable for them through the contractor. The only way the employer can obtain any contractual redress is if he has entered into the RIBA/CASEC form, though he could sue the subcontractor for negligence. Stranger still, no other subcontractor is to be responsible for them through the contractor and, in that case, there is no way in which the employer can obtain redress, since there is no privity of contract between the employer and the contractor's ordinary subcontractor under clause 3.2. The phrase in this clause is that neither the contractor nor the named person, or any other subcontractor, is responsible under the contract 'for anything to which the above terms relate'. The precise meaning of this phrase may be a fruitful source of dispute. The clause, however, does not affect the contractor's normal obligations or those of any subcontractor in the supply of goods and materials and workmanship.

There are extensive provisions to deal with the determination of the named person's employment under the subcontract and its consequences.

The contractor must advise you of any events which are likely to lead to determination under the subcontract. He must do this 'as soon as is reasonably practicable'; in other words, he must let you know as soon as he finds out. What you are to do then is not stated. Presumably, you will endeavour to find a solution short of determination.

If determination takes place, whether or not you have been advised of the likelihood, the contractor must write to you, giving the circumstances. You are then to issue such an instruction as may be necessary, which must be one of the following:

— name another person to do the work or the balance of the work,

Subcontractors and suppliers

incorporating a description and particulars of the named person in sections I and II of Form of Tender and Agreement NAM/T, subject to the contractor's reasonable objection within 14 days as in clause 3.3.2(c) (clause 3.3.3(a))

— instruct the contractor to make his own arrangements to do the work, either himself or by subcontract under clause 3.2 (clause 3.3.3(b))

— omit the work still to be finished (clause 3.3.3(c)).

If you omit the work, the employer may, under clause 3.11, arrange to have the work done by others and pay direct.

The contract points out a difference in consequence, depending on whether the work was originally included in the contract documents or is the result of an instruction regarding the expenditure of a provisional sum.

— if work originally included in the contract documents: a clause 3.3.3(a) instruction ranks as a clause 2.3 event (extension of time) but not as a clause 4.10 matter (disturbance of progress). The contract sum must be increased or reduced to take account of the price of the second as compared with the first named person. Amounts included in the second price for repair of the first named person's defective work must be excluded from the contract sum. Thus, the contractor is not to be held financially responsible for the determination, but he is held responsible for any defects at that time existing. A clause 3.3.3(b) or (c) instruction ranks as a clause 2.3 event (extension of time) *and* as a clause 4.10 matter (disturbance of progress). It also ranks as a variation.

— if the result of an instruction regarding the expenditure of a provisional sum: a clause 3.3.3(a), (b), or (c) instruction ranks as a clause 2.3 event (extension of time) and as a clause 4.10 matter (disturbance of progress). It also ranks for payment as a further instruction under the provisional sum.

There is an important proviso that, if the instruction was issued as a result of some default of the contractor, none of the above *benefits* apply to the contractor. This is so whether the work was originally included in the contract documents or is the result of an instruction regarding the expenditure of a provisional sum. It is only fair to stress that the above is intended to clarify what is regarded as an unhappily worded clause. A strict reading of clause 3.3.5, for example, makes the employer liable to pay the contractor for the subcontract work if the contractor is in default and a clause 3.3.3(c) instruction omitting the work is issued. That cannot be what was intended.

The contractor must take whatever action is necessary, within reason, to recover from the named subcontractor any additional amount that the employer has had to pay to the contractor due to the issue of the

8.2 Subcontractors

instructions 3.3.3(a), (b) or (c) and the consequences thereof; and any liquidated damages which the employer would have been able to recover from the contractor had it not been for the instructions. There is no time limit set for the contractor to take the action. Presumably, he will take the action after practical completion; otherwise, how is he to know the amount of liquidated damages which would have been recoverable by the employer? (Clause 3.3.6)

The contractor is required to take the action only where he has determined the employment of the subcontractor under NAM/SC clauses 27.1 or 27.2. There is a proviso that he cannot be required to commence any arbitration or other proceedings unless the employer agrees to pay any legal costs he incurs. It is difficult to see what action the contractor can take (unless he is holding some of the named subcontractor's money) if the employer decides against paying his legal costs. The contractor is to account—ie, presumably makes an account—for any amounts he recovers. Insofar as the contractor fails to take action, he is liable to the employer for the additional amount, including the amount equal to liquidated damages, payable by, or due to, the employer. Whether or not the employer agrees to fund legal proceedings, it is difficult to understand how you are to demonstrate, even to your own satisfaction, that the contractor has failed to take action, provided he has written some sternly worded letters to the named person. We do not consider that, taken as a whole, the provisions for named persons are satisfactory at the moment. No doubt they will be revised.

8.3 Statutory authorities

Certain crucial parts of most, if not all, contracts are carried out by local authorities or statutory undertakers such as the gas and electricity boards. Where they carry out the work solely as a result of their statutory rights or obligations, they are not to be considered named persons (clause 3.3.8). The implication is that, where they carry out work which is not a result of their statutory rights or duties, but as a matter of contract, they may be named persons. They may also, in such instances, be considered as subcontractors to the contractor in the traditional way (clause 3.2) or persons employed by the employer to carry out work outside the contract (clause 3.11; see section 8.4). In carrying out their statutory duties, the authorities have no contractual liability, although in some cases they have tortious liability. When they are carrying out work outside their statutory duties, they are exactly like anyone else who enters into a contract. If an authority delays the completion of the works by carrying out (or failing to carry out) work in pursuance of its statutory duty, the contractor will be entitled to an

Subcontractors and suppliers

extension of time (clause 2.4.13), but not for delay caused by other work which the authority carries out or fails to carry out.

The contractor must comply with all statutory requirements: for example, the Planning Acts and dependent regulations. He is responsible for giving any notices required and for paying any fees or charges in connection with the works (clause 5.1), for which he is entitled to be reimbursed by having the amounts added to the contract sum, unless they are already included in the contract documents. Thus, if the appropriate contract document is a specification and it states that the contractor must allow for paying all statutory fees and charges, he will be deemed to have included the amount in his price.

The contractor is not liable to the employer if the works do not comply with statutory requirements if this is because he has carried them out in accordance with the contract documents of any of your instructions (clause 5.3). There is a proviso that, if the contractor finds any divergence between the statutory requirements and the contract documents or your instructions, he must specify the divergence to you in writing (clause 5.2). It may be small comfort, because, if he does not find a divergence which exists, he is not liable. Of course, you can always argue that he should find all divergences as part of his general obligations, but you may have difficulty convincing an arbitrator of this.

If an emergency arises, the contractor must comply with statutory requirements without waiting for your instructions. He must carry out and supply just enough work and materials as are necessary to comply as an immediate measure, and inform you forthwith, not necessarily in writing. If he satisfies these requirements, he is entitled to be paid as though he has carried out your instructions requiring a variation, provided that the emergency arose because of a divergence between the statutory requirements and the contract documents and/or any of your instructions, drawings, or documents issued under clauses 1.3, 3.5, or 3.9 (clause 5.4).

Thus, if the emergency arises through the contractor's default or inefficiency, he is entitled to nothing except insofar as you may be prepared to accept that there would have been some additional cost irrespective of the contractor's default.

8.4 Work not forming part of the contract

The employer has the right to enter into contracts with persons other than the contractor to carry out work on the site (clause 3.11). Such persons are commonly firms or individuals over whom the employer wants complete control. They may be his own employees as, for example, when the employer is a local authority. Or the employer's reason for wanting to employ such persons in a direct way (ie, not

8.4 Work not forming part of the contract

through the contractor) is that they have a special relationship with him: they may be artists, sculptors, graphic designers, landscapers, etc.
In principle, it is always wise to make the contractor responsible for all the work to be done. It promotes efficiency on site and removes areas of possible dispute. If the employer insists on having some directly employed persons, remember that, for work to be considered not to form part of the contract, it must be the subject of a separate contract between the employer and the person to provide the work, and it must be paid for by the employer directly to the person employed and not through the contractor.

Work carried out by statutory authorities not in pursuance of their statutory duties may fall into this category.

The contract provides for two situations. The first is where the contract documents provide for such work by informing the contractor what is to be carried out, when, and (possibly but not necessarily) by whom. The documents should require the contractor to allow the work to be carried out as stated and give details of any items of attendance which may be required. The contractor should then make provision for the work in his programme. The second situation is where the contract documents do not provide for such work.

In the first situation, the contractor must allow the work to be carried out and provide whatever attendance is specified and priced for. In the second, the employer must first obtain the contractor's consent to his proposals before arranging to have the work carried out. The contractor must not withhold his consent unreasonably. It would be reasonable to withhold consent if the proposed work would constitute a severe disturbance to the contractor's progress. It would be unreasonable if the work was to be done without affecting the contractor's activities in any way. Between these two extremes lie many situations which are not easy to resolve.

Whether the work not forming part of the contract is provided for in the contract documents or is simply the subject of the contractor's consent does not affect the employer's responsibility under the contract. For the purposes of the insurance clause 6, persons directly employed by the employer are deemed to be persons 'for whom the employer is responsible'. They are not to be deemed subcontractors. The result is that the employer may have uninsured liabilities. The best way to rectify this is to advise the employer to obtain the necessary cover through his insurance broker. The broker should be given a copy of the insurance clauses and be requested to arrange cover for the employer, and those for whom the employer is responsible, in respect of any act or neglect of those persons. The directly employed persons may already have adequate insurance cover, but it is a matter best left in the hands of a broker with experience in this kind of insurance.

Subcontractors and suppliers

The contractor is entitled to be awarded extensions of time if completion is delayed by the carrying out (or failure to carry out) of work not forming part of the contract (clause 2.4.8). It is a very easy claim to make, and a contractor running late on the contract would do well to give his consent, if required, to any work by the employer's directly-employed persons. They can then be made to appear responsible for any delays thereafter. You must, therefore, be especially vigilant when examining such claims; a network analysis is invaluable.

The employer's responsibility to supply those materials which he has agreed to supply, or his failure to do so, is closely linked to the problem of work outside the contract (clause 2.4.9). In both cases, the contractor is not involved with the external contracts, and it may be thought that the employer is assuming needless responsibility. Suppose that the employer wishes to supply all the paint for a contract himself, perhaps because he thinks he can obtain it at a cheap rate. In order to avoid any claims from the contractor, he must supply paint of the correct colours, in the correct quantities, of the correct types (undercoat, etc), and at the time it is required. Moreover, any unsatisfactory paint may provoke a claim.

The wording of the clauses relating to extensions of time, discussed above, is precisely repeated as matters for which the contractor may make a claim for loss and/or expense (clauses 4.12.3 and 4.12.4). The employer is extremely vulnerable, and you must exercise great care in deciding whether the claim is valid.

Potentially the most damaging result of the employer's directly employing persons to carry out work outside the contract, or of his arranging to supply materials himself, is that the contractor may acquire grounds to determine his employment (clause 7.5.3(c)). The matter is dealt with in section 12.2.2.

If the employer expresses his intention of employing his own men or of supplying any materials, you have a duty to advise him of the pitfalls. Most of them can be avoided by ensuring that all work is done, and all materials are supplied, through the contractor. In view of the possibility of named persons as subcontractors in this contract, in our opinion the employer would be unwise to arrange for work to be carried out which does not form part of the contract.

8.5 Summary

Assignment and subcontracting

— neither party may assign without the other's consent
— the contractor may sublet with your consent

8.5 Summary

— the subcontractor's employment determines on the determination of the contractor's employment
— the contractor must consent to the subcontractor's removing unfixed materials from site
— unfixed materials become the employer's property when he pays the contractor
— unfixed materials become the contractor's property if he pays the subcontractor before being paid himself.

Named persons

— named persons may arise by inclusion in the contract documents or by an instruction regarding a provisional sum
— sections I, II, and III of tender and agreement NAM/T must be completed by you, the named person, and the contractor
— the work may be omitted and carried out by the employer's own men
— the named person in the contract documents may be replaced by a named person in your instruction
— the contractor may object to any named person in an instruction
— the employer should enter into the RIBA/CASEC form of agreement with the named person
— design, selection, and satisfaction are not the liability of the contractor, nor are they the liability of the named person or any subcontractor through the contractor
— on determination of a named person's employment, you may name another person, request the contractor to make his own arrangements, or omit the work
— under certain circumstances, extension of time and loss and/or expense may be awarded to the contractor after determination
— the contractor must take action to recover the employer's losses after determination of the named person's employment
— if the contractor fails to take action, he is liable for the losses.

Statutory authorities

— in pursuance of their statutory duties, statutory authorities are not liable in contract but may give grounds for extension of time
— not in pursuance of their statutory duties, they are liable in contract and may be subcontractors, named persons, or persons for whom the employer is responsible
— the contractor must comply with statutory requirements
— the contractor is not liable to the employer if he works to

Subcontractors and suppliers

contract documents or instructions, provided that he notifies any divergence he finds
— the contractor may carry out and be paid for emergency work without instruction
— the contractor must notify you forthwith if there is an emergency.

Work not forming part of the contract

— the employer may carry out his own work on site
— the contractor's consent is required if the work is not mentioned in the contract documents
— there are insurance implications
— the employer is vulnerable to claims for extension of time and loss and/or expense
— there are grounds for determination if work is delayed for a month by work not forming part of the contract.

9 Possession, practical completion, and defects liability

9.1 Possession

9.1.1 General

Possession is the next best thing to ownership. If you lend your motorcar to a friend, he has a better claim to the car than anyone else except you. The builder in possession of a site can, in general terms, exclude everyone from the site except (and often including) the owner. In practice, there are exceptions to this general rule, laid down by the building contract and by various statutory regulations.

A contractor carrying out building works is said to have a licence from the owner to occupy the site for the length of time necessary to complete the work. The owner has no general power to revoke such a licence during the contract period, but it may be brought to an end if the contractor's employment or the contract itself is lawfully brought to an end.

If there were no express term in the contract giving the contractor possession of the site, a term would be implied that the contractor must have possession in sufficient time to allow him to complete by the contract completion date.

There is no express term in IFC84 allowing access to anyone other than the contractor. So a term must be implied to allow access to the works at all reasonable times for the architect, clerk of works, and all properly appointed subcontractors. (Curiously, the subcontract form NAM/T does provide for the contractor, the architect, and all persons duly authorised by them to have right of access to work being prepared in the subcontract works [NAM/T clause 23]).

Possession, practical completion, and defects liability

9.1.2 Date for possession

The date for possession is to be entered in the appendix. Clause 2.1 states that, on that date, possession must be given to the contractor, and that, subject to any extension of time which may be awarded, the contractor must proceed regularly and diligently with the works and complete them on or before the date for completion in the appendix.

If the employer fails to give sufficient possession on the due date, it is normally a serious breach of contract; the contractor will have a claim for damages at common law, and the time for completion will become 'at large'. That is to say, the contractor's obligation will be to complete the work within a reasonable time, and no date can be established from which liquidated and ascertained damages begin to run. The problem is partially overcome in IFC84 by the inclusion of an optional clause (2.2), which permits the employer to defer giving possession for a time which must not exceed a period to be inserted in the appendix. You would be well advised to ensure that the clause is included for the employer's protection. The contract recommends that the period of deferment should not exceed six weeks. The period is somewhat arbitrary because, at the time of tender—of signing the contract even—presumably the employer fully intends to give possession on the date stated in the appendix. If he knew, at that time, that the date was going to be deferred, the date in the contract could be adjusted accordingly. It is perfectly possible to amend the contract so that the permitted deferment is, say, 12 weeks, but to introduce so large an element of uncertainty into the contract would almost certainly result in increased tender prices. Possession is usually delayed for a very short time (because demolition contractors have not finished their work, or because planning or building-regulation permission is delayed, or for some other similar cause). Otherwise, the delay is caused by some major problem which lasts for a considerable period, and in such circumstances it would be unfair to rely on a clause permitting deferment. The best-laid plans can go wrong, as any architect knows, so although you may, in consultation with the employer, decide to insert a period shorter than six weeks, do not omit an insertion altogether.

There is no prescribed form of notice for deferment, but it must come from the employer. You should draft him a suitable letter, which need not give any reason, Fig 9.1.

If possession is deferred, the contractor will be able to claim loss and/or expense (clause 4.11(a)) and an extension of time (clause 2.4.14). The power to defer possession must not be confused with your power to order postponement of the work to be executed under the provisions of the contract. Loss and/or expense (clause 4.12.5) and extension of time

9.1 Possession

Fig 9.1
Employer to contractor, deferring possession of the site

Dear Sir

[*If length of deferment is known*]

In accordance with clause 2.2 of the conditions of contract, take this as notice that I defer giving possession of the site for [*specify period*]. You may take possession of the site on [*insert date*].

[*If length of deferment is not known*]

In accordance with clause 2.2 of the conditions of contract, take this as notice that I defer giving possession of the site for a period not exceeding [*insert the period named in the appendix*]. I will write to you again as soon as I have a definite date for you to take possession.

Yours faithfully

Copy: Architect
      Quantity surveyor
      Consultants
      Clerk of works

(clause 2.4.5) can also be claimed for postponement. If the whole, or substantially the whole, of the works is suspended for one month by postponement, the contractor may determine his employment (clause 7.5.3). If there is postponement, the contractor has possession of the site, but you have suspended work. That is not to say that the contractor may not use some or all of the time the work is suspended to work specifically connected with his occupation of the site (for example, repairing or improving site office accommodation, sorting materials, attending to security).

The contractor normally gives up possession of the site at practical completion or on determination of this employment. He is, however, granted a restricted licence to enter the site for the purposes of remedying defects (see section 9.3).

9.2 Practical completion

9.2.1 Definition

Clause 2.9 states that you must issue a certificate forthwith when, in your opinion, practical completion of the works is achieved. The consequences of your certificate are considerable (see section 9.2.2). Despite that, the contract does not define the meaning of 'practical completion' (although it takes the trouble to define 'person' [clause 8.3]). It is not the same as 'substantial completion', nor does it mean 'almost complete'. There is conflicting case law, but the point that emerges seems to be that you are not to certify practical completion if any defects are apparent or if anything other than very trifling items remain outstanding. Within these guidelines, you are free to exercise your discretion. If some items are outstanding and the contractor is pressing you to issue your certificate, ask yourself whether the employer is going to be seriously inconvenienced while these items are being finished. If he is, then you would be justified in withholding your certificate.

There is no obligation on you to tell the contractor what items remain to be completed. The temptation to issue lists of outstanding items should be resisted. The contractor knows what is required of him by the contract. The issue of lists at this stage is confusing and often leads to disputes. The onus of inspecting the work and preparing work lists for the contractor lies with the person-in-charge.

The contractor is not bound to notify you when practical completion has been achieved, but he is certain to do so, probably some weeks in advance. Some architects are in the habit of arranging so-called 'handover meetings' at which representatives of the employer and sometimes his maintenance organisation are present. This can be a

9.2 Practical completion

prudent move on the principle that many eyes are better than one. You will also have any consultants on hand to inspect their own particular portions of the work. Remember, however, that the decision to issue a certificate is yours. You cannot move the responsibility on to the employer simply because he is present. The exception to that is if the employer insists that the building is ready, even though you make clear to him your dissatisfaction. He may agree with the contractor to take possession of the building despite your protests. This often happens if an employer is anxious to get into a building, but you should not be tempted to issue your certificate. There is no contractual requirement for you to do so. Write to the employer and make the position clear (Fig 9.2). If the employer later discovers that outstanding items cause him trouble, or if the contractor does not complete as quickly and efficiently as he promised, the employer will have only himself to blame. You will have carried out your duties properly with due regard for his interests. Your duty to issue a certificate of practical completion will remain, but not until, in your opinion, the works have achieved that state.

9.2.2 Consequences

Clause 2.9 states that 'Practical Completion of the Works shall be deemed for all the purposes of this contract to have taken place on the day named in' the certificate. The 'purposes of this contract' are to be found in a number of clauses throughout the contract. They are as follows:
— the contractor's liability for insurance under clause 6.3A ends
— liability for liquidated damages under clauses 2.7 ends
— liability for frost damage ends (clause 2.10)
— the employer's right to deduct full retention ends. Half the retention percentage becomes due for release within 14 days (clause 4.3)
— the period of final measurement and valuation begins (clause 4.5)
— the period of final review of extensions of time begins (clause 2.3)
— the defects liability period begins (clause 2.10)

All these 'purposes', with the possible exception of the review of extensions of time, are positively beneficial to the contractor and are ample reason for him to be anxious to secure a certificate of practical completion at as early a date as possible.

9.3 Defects liability period

9.3.1 Definition

Clause 2.10 refers to defects liability. Reference is also made to the defects liability period, which is to be 'named in the Appendix'. The

Fig 9.2
Architect to employer if employer wishes to take possession of building before practical completion achieved

Dear Sir

I refer to our meeting on site with the contractor on [*insert date*].

I confirm that, in my opinion, practical completion has not been achieved, so it is my duty under the provisions of the contract to withhold my certificate. I note, however, that you have agreed with the contractor to take possession of the building for reasons of your own. Although I think that you are unwise, as I explained on site, I respect your decision and I will continue to inspect until I feel able to issue my certificate. At that date, the defects liability period will commence.

Yours faithfully

9.3 Defects liability period

period is for the benefit of all parties but principally for that of the contractor. The idea is to allow a specific period of time for defects to appear, list the defects, and give the contractor the opportunity to remedy them. Any defect is a breach of contract on the part of the contractor, who has agreed to carry out the work in accordance with the contract documents. If there were no defects liability period, the employer's only remedy for defects would be to take action at common law. So the insertion of the period provides a valuable method of identifying defective work and having it corrected. Without it, the contractor would have no right or duty to return.

It is important to remember that the contractor's liability for defects does not end at the end of the defects liability period; what does end is his right to correct them (and even that right is limited, as will be seen). Afterwards, the employer is free to take legal action for damages if further defects appear, although, in practice, the employer will normally be satisfied if the defects are corrected.

Contractors commonly refer to the defects liability period as the 'maintenance period'. This is misleading and wrong. Maintenance implies a far greater responsibility than simply making good defects—for example, touching up scuffed paintwork and attention to general wear and tear.

9.3.2 Defects, shrinkages, or other faults

The contractor is required to make good 'defects, shrinkages, or other faults'. At first sight, this might appear to be all-embracing. In fact, 'other faults' is to be interpreted *ejusdem generis*. That is to say that they must be faults which are similar to defects or shrinkages. So a defect occurs when something is not in accordance with the contract. If an item of workmanship is in accordance with the contract, it is not defective for the purposes of this clause. It might be less than adequate in some way, but that could be due to a fault in design, and so your responsibility. Shrinkages are a source of dispute on many contracts. They become the contractor's liability only if they are due to materials or workmanship not in accordance with the contract.

For example, shrinkage most commonly occurs in timber, caused by a reduction in moisture content after the building is heated. It is your job to specify a suitable moisture content for the situation. If shrinkage occurs during the defects liability period, it can only be because the timber was supplied with too high a moisture content, or because your assumption about the appropriate moisture content was incorrect. Only the first explanation is the contractor's liability. In practice, it is often very difficult to decide which explanation applies, and you will naturally be drawn to the conclusion that your specification is correct.

115

If the contractor objects, the only way to be sure is to have samples cut out and tested in the laboratory, a very expensive procedure. All too often, the culprit is the occupier of the building, who is running the central heating above the recommended temperatures. In such a case, your assumptions were, in effect, wrong, and the contractor is not responsible for making good.

9.3.3 Frost

The contractor's liability to make good frost damage is limited to damage caused by frost which occurred before practical completion—in other words, when the contractor was in control of the building works and could have taken appropriate measures to prevent the damage by introducing heating or stopping vulnerable work. Any damage caused by frost after practical completion is at the employer's own expense. The difference is usually easy to spot on site. Frost damage occurring after practical completion is often due to faulty detailing or maintenance.

9.3.4 Procedure

The defects liability period starts on the date given in the certificate of practical completion as that on which practical completion was achieved. You are to fill in the length of period required in the appendix. If you do not fill in any period, the length will be six months. You should agree the period with the employer, who will ask for your advice. Although six months is a common defects liability period, there are really no good reasons why the period should not be extended to nine or, better, 12 months. Specialist work such as heating often has a 12-month period to fully test the system through all the seasons of the year. It is probable that a contractor asked to tender for a contract including a 12-month general defects liability period would increase his tender figure slightly, but lengthening the period does not increase his actual liability, but only his right to return to make good defects. The increase in tender price probably reflects the confusion with 'maintenance period' and the fact that the final certificate would be delayed somewhat. In fact, only 2½ per cent of the contract sum should be outstanding after practical completion (clause 4.3).

The equivalent clause (17.1) in JCT80 refers to defects, etc, which 'shall appear within' the period. In IFC84, the clause simply refers to defects, etc, 'which appear'. Although the time limit is the end of the period, the wording of the clause seems to give you the power to notify the contractor of all defects which appear, including those which are present at the date of practical completion. Some commentators

9.3 Defects liability period

maintain that the JCT80 clause gives you that power anyway, but it is by no means certain. IFC84 removes any uncertainty.

You are to notify defects to the contractor not later than 14 days after the expiry of the period. This is normally done as soon as possible after the end of the period. You will have inspected just before the period expired. There is no set form for notifying the contractor, but it is advisable to send him a letter (Fig 9.3) enclosing a schedule of defects. Your power to require defects to be made good is not confined to the issue of the schedule. The wording of the clause makes it clear that you can notify the contractor at any time within the period.

The requirement is for the contractor to make good the defects which you notify to him. No particular time limit is set, but he must carry out his obligation within a reasonable time. What is a reasonable time will depend on the circumstances, including the number of defects, their type, and any special arrangement to be made with the employer for access.

Ideally, the contractor should return to site within a week or so after receiving your list and bring sufficient labour to make good the defects within, say, a month. If you decide to exercise your right to require defects to be made good during the currency of the period, you should confine such requests to really urgent matters to be fair to all parties.

All defects which are the fault of the contractor are to be made good at his own cost. However, there is an important proviso. If the employer agrees, you may instruct the contractor in a letter (Fig 9.4) not to make good some or all of the defects, and an appropriate deduction must be made from the contract sum. The clause gives no guidance on the method of arriving at an appropriate deduction. The job is best left to the quantity surveyor. Although no provision is made for the contractor to agree the amount of any such reduction, obviously it must be based on the cost of making the defects good. Bear in mind that the contractor would be able to carry out the work at less cost than, say, another contractor brought in for the purpose. Although you do not have to give the contractor a reason for not requiring him to make good, you will probably issue such instructions only if the employer prefers to live with the defects rather than suffer the inconvenience of the contractor returning, or if the contractor's work record is so bad during the carrying out of the contract that you have no confidence that he will make a satisfactory job of making good. In the second instance, you would probably be justified in obtaining competitive quotations for the work and deducting the amount from the contract sum. Make sure that you obtain a letter from the employer authorising you to instruct the contractor that making good is not required (letter, Fig 9.5).

When you are satisfied that the contractor has properly completed all making good of defects notified to him, you must issue a certificate to

Fig 9.3
Architect to contractor, enclosing schedule of defects

Dear Sir

The defects liability period ended on the [*insert date*]. I inspected the works on the [*insert date*]. In accordance with clause 2.10 of the conditions of contract, I enclose a schedule of defects. I should be pleased if you would give them your immediate attention.

Yours faithfully

Copy: Clerk of works

9.3 Defects liability period

Fig 9.4
Architect to contractor, instruction him not to make good

Dear Sir

The defects liability period ended on [*insert date*]. I inspected the works on [*insert date*]. I enclose a schedule of the defects I found.

[*Then either*]:
I hereby instruct that, in accordance with clause 2.10, you are not required to make good any of the defects shown in the schedule.

[*Or*]:
I hereby instruct that, in accordance with clause 2.10, you are not reqired to make good those defects marked 'E'.

[*Then*]:
An appropriate deduction will be made from the contract sum in respect of the defects which you are not required to make good.

Yours faithfully

Copy: Employer
      Quantity surveyor
      Clerk of works

Fig 9.5
Architect to employer if some defects are not to be made good by contractor

Dear Sir

Further to our conversation/your letter/my letter [*delete as appropriate*] of [*insert date*], I enclose a list of defects found at the end of the defects liability period. I understand that you agree that the contractor should be instructed not to make good the defects which I have marked 'E'/any of these defects [*delete as appropriate*]. An appropriate deduction will be made from the contract sum.

In order that I may issue my instruction in accordance with the contract, I should be pleased to have your written consent to the course of action outlined above.

Yours faithfully

that effect. The certificate is important because it marks one of the dates starting the 28-day period within which the final certificate must be issued (clause 4.6).

9.4 Summary

Possession

— possession must be given on the date shown in the appendix but—
— it may be deferred for a maximum period stated in the appendix, not exceeding six weeks
— deferment gives grounds for extension of time and a claim for loss and/or expense
— possession ends at practical completion.

Practical completion

— practical completion requires your certificate
— it is a matter for your opinion alone
— it marks the date on which many of the contractor's obligations end.

Defects liability period

— the defects liability period is for the benefit of the contractor
— the end of the period does not mark the end of the contractor's liability for defects
— defects can only be workmanship or materials not in accordance with the contract, or damage caused by frost occuring before practical completion
— the length must be entered in the appendix
— all defects apparent during the period or on practical completion are covered
— defects must be notified to the contractor during the period and not later than 14 days after the end
— defects are to be made good at the contractor's own cost, or—
— if the employer agrees, you may instruct the contractor not to make good defects and deduct an appropriate sum from the contract sum
— after the contractor has made good, you must issue a certificate to that effect.

10 Claims

10.1 General

This chapter covers claims by the contractor for both extra time and extra money, though there is not necessarily any link between the two. The subject of claims is an emotive one, but as architect you have a duty under the contract to deal with claims, and in so doing you must hold the balance fairly between the employer and the contractor.

Two points should be noted. The first is that your powers under the contract are limited. You can settle only claims which you are authorised to deal with under the express terms of the contract. This means that you cannot deal with common-law claims or make *ex gratia* settlements. To do that, you need the express authority of the employer. Table 10.1 summarises the contract clauses which give rise to claims which you are empowered to settle and those which are the employer's province.

The second point is that if you reject a valid claim under the contract, case law suggests that this is a breach of contract for which the employer is responsible at common law. It is also clear (since *Sutcliffe* v *Thackrah* [1974]) that potentially you can be liable to the contractor for any loss or damage which he suffers or incurs as a result of your negligence: eg, if for any reason he cannot recover against the employer. These observations apply to claims for extra money under the contract provisions or claims for 'direct loss and/or expense', as the contract puts it. There is an important point to note about extensions of time: failure by you properly to exercise your duties as to the granting of extensions of time may result in the contract completion date becoming unenforceable, in the sense that the contractor will no longer be bound

10.1 General

Table 10.1
IFC84 clauses which may give rise to claims (*see key on page 124*)

Clause	Event	Type	Usually dealt with by
1.4	Inconsistencies, etc	C	A
1.6	Architect fails to provide documents	CL	E
1.8	Divulging or improper use of rates	CL	E
2.2	Deferment of possession Deferment exceeding time stated in appendix	C CL	A E
2.3	Financial claims	CL	E
2.7	Improper deduction of liquidated damages	CL	E
2.8	Failure to repay liquidated damages	CL	E
2.9	Failure to issue certificate of practical completion	CL	E
2.10	Including items which are not defects Failure to issue certificate of making good defects	CL CL	E E
3.1	Assignment without consent	CL	E
3.2	Architect unreasonably withholding consent to subletting	CL	E
3.3.1	Failure to issue instruction	CL	E
3.3.3	Failure to issue instruction	CL	E
3.3.3(b)	Contractor instructed to make own arrangements after determination	C	A
3.3.7	Failure of subcontract design, selection of materials, or satisfaction of performance specification	CL	E
3.5.1	Instructions issued orally or by employer	CL	E
3.6	Valuation of variations not carried out in accordance with the contract	CL	E
3.9	Failure to determine levels, etc	CL	E
3.10	Clerk of works exceeding duties	CL	E
3.11	Employer's work Employer's work without going through procedure	C CL	A E

Claims

Table 10.1
IFC84 clauses which may give rise to claims (*see key below*)—*cont.*

Clause	Event	Type	Usually dealt with by
3.12	Opening up and testing	C	A
3.13.2	Unreasonable instructions following failure of work	C	Arb
3.14	Wrongly phrased instructions	CL	E
3.15	Postponement of work	C	A
4.2	Certificate not issued or not issued at the proper time	CL	E
4.3	Failure to certify	CL	E
4.4	Retention money, interest	CL	E
4.5	QS's failure to send computations within period of final measurement	CL	E
4.6	Failure to issue final certificate, or certificate not in proper form	CL	E
4.9, 4.10	Fluctuations	C	A
4.11	Disturbance of regular progress	C	A
5.1	Statutory fees and charges	C	A
5.2	Divergence between statutory requirements and contract documents	C	A
5.4	Emergency compliance	C	A
5.5	Recovery of VAT	C	A
6.3A.4	Failure to pay insurance monies	CL	E
6.3B.2	Employer's failure to insure	C	A
6.3C.1	Employer's failure to insure	C	A
7.1	Invalid determination	CL	E
7.8.1	Invalid determination	CL	E

KEY
C–Contractual; CL–Common Law; E–Employer; A–Architect; Arb–Arbitrator

10.2 Extension of time

by the contract period, but will merely be required to complete 'within a reasonable time'. The important consequence of this is that the employer would forfeit any right to recover liquidated damages, which would leave him with the difficult and unenviable task of proving his actual loss at common law.

Great care is therefore needed in dealing with claims for both time and money.

10.2 Extension of time

10.2.1 Legal principles

Under the general law, the contractor is bound to complete the works by the agreed date, unless he is prevented from so doing by the employer's fault or breaches of contract—and the employer's liability extends to your wrongful acts or defaults within the scope of your authority. Unless there is an extension-of-time clause in a contract, neither you nor the employer have any power to extend the contract period.

Clause 2.3 deals with your power to grant extensions of time and lays down the procedure which must be followed. It is closely linked with clause 2.7—which provides for liquidated damages—and case law establishes that clauses in this form will be interpreted very strictly by the courts. In practice, this means that if delay is caused by something not covered by the clause—or if you fail to exercise your duties under it properly and at the right time—the employer will lose his right to liquidated damages.

10.2.2 Liquidated damages

Clause 2.7 provides for the contractor to 'pay or allow' the employer liquidated damages at the rate specified in the appendix should he fail to complete on time. You should have calculated the amount of liquidated damages carefully at pre-tender stage. The figure must represent a genuine pre-estimate of the loss likely to be suffered by the employer should the contractor fail to complete on time.

To pitch the figure at the right level, you should discuss it carefully with your client and then make a calculation. If the sum arrived at is a genuine pre-estimate of the likely loss, than that is the sum which will be recoverable, even if the sum agreed is greater than the actual loss or there is no loss.

The calculation is sometimes difficult. In the case of profit-earning assets, there is no problem, and all you need to do is analyse the likely losses and additional costs. The following should be considered:

Claims

— loss of profit on a new building: eg, rental income, retail profit
— additional supervision and administrative costs
— any other financial results of the contract's being late: eg, staff costs
— on-costs under later direct contracts: eg, a contract for fitting out
— interest payable during the delay.

These points are not exhaustive, and clearly much depends on the type of project. For example, in the public sector, late completion may require the temporary occupation of a more expensive building, and invariably there will be extra administrative costs and almost inevitably some financial penalties for late completion.

There are three possible bases for formulating liquidated damages in such cases:

1 You can take the notional rental value of the property and base on this some notional return on its capitalised value. We do not recommend this option, nor its variant of fixing a sum commensurate with an average commercial project of comparable value.

2 You can use the sources of likely loss as a basis—for instance, subsidies, capitalised interest on money advanced, and the spin-off effect on departmental costs. This method normally produces a nominal figure which does not encourage the tardy contractor to make up lost time. If the figure is unrealistic and the contractor has two contracts running late, he will be tempted to throw his resources into the contract which has the more realistic provision for liquidated damages.

3 The best method is to use a formula calculation that gives an approximation of all individual costs. The best known is that put forward by the Society of Chief Quantity Surveyors in Local Government. This suggests three main headings under which a calculation should be made:

— assume that 80 per cent of the total capital cost of the scheme (including fees) will have been advanced at anticipated completion and that interest is being paid at the current rate. If that interest rate is 12 per cent, capitalised interest is 80 per cent × 12 per cent ÷ 52 = 0.185 of the contract sum per week
— assess administrative costs (eg, staff salaries) as 2.75 per cent of the contract sum per year. This gives 0.053 per cent per week
— exceptional costs (eg, temporary accommodation) are assessed realistically.

The result of the first two parts of the formula reduces to 0.237 per cent of the contract sum per week. This may be an underestimate of the likely losses, but it works reasonably in practice.

The main point is that the figure should not be plucked out of the air. You must make a genuine attempt to calculate the likely loss and use

10.2 Extension of time

the resultant figure. In commercial projects, the likely loss may produce a figure out of all proportion to the value of the contract. The solution then is to reduce the figure to an acceptable level in consultation with the employer.

Fig 10.1 is a possible format for a general calculation for liquidated damages.

There are two pre-conditions for the deduction of liquidated damages under clause 2.7.

The first is the issue of your certificate of non-completion, stating that the contractor has failed to complete the works by the original or extended date for completion. This certificate is a factual statement and you can issue more than one. Should you make an extension of time after issuing a certificate of non-completion, the contract says that you must issue a written cancellation of it and issue a further certificate as necessary.

Figs 10.2 and 10.3 are pro forma examples.

The second is that the employer must require the contractor to pay or allow liquidated damages. He must do this by writing to the contractor to that effect no later than the date of the final certificate for payment (see clause 4.6).

If you cancel your certificate of non-completion because you have extended time, any liquidated damages deducted by the employer in the meanwhile must be repaid by him to the contractor. For example, you have issued a certificate of non-completion and the employer has deducted four weeks' liquidated damages at £750 a week, making a total of £3000. You then issue a variation instruction, and assess the resultant delay at three weeks, cancelling your first certificate and issuing a new one once the extended date is passed. The employer must then repay 3 × £750 = £2250 to the contractor. The employer repays this sum net; the contractor has no right to interest on the money, either under the contract or under the general law (clause 2.8).

10.2.3 Procedure

Clause 2.3 is basically a simplified version of JCT80, clause 25, but the provisions for the contractor to notify you of delays affecting progress are less detailed, and there are other important differences.

The flowchart Fig 10.4 illustrates the contractor's duties in claiming an extension of time under clause 2.3.

The flowchart Fig 10.5 sets out your duties in relation to such a claim.

The procedure under clause 2.3 is as follows:

As soon as progress is, or is likely to be, delayed—the contract says 'upon it becoming reasonably apparent'—the contractor must give you written notice of the cause of the delay *forthwith*. The wording of clause

Fig 10.1
Calculation of liquidated and ascertained damages (typical format)

Contract . . . .
Client   . . . .
Architect . . . .

	costs/week
1 SUPERVISORY STAFF (current rates)	
Architect:	
Estimated hrs/wk ... x time charge of £.../hr	£
Quantity surveyor	
Estimated hrs/wk ... x time charge of £.../hr	£
Consultants [*as above for each one*]	£
Clerk of works	
Weekly salary (= yearly ÷ 52)	£
Total (1)	£
2 ADDITIONAL COSTS (current rates)*	
Rent and/or rates and/or charges for present premises	£
Rent and/or rates and/or charges for alternative premises	£
Charges for equipment	£
Movement of equipment	£
Additional and/or continuing and/or substitute staff	£
Movement of staff (include travel expenses)	£
Any site charges which are the responsibility of the client	£
Extra payments to directly employed trades	£
Insurance	£
Additional administrative costs	£
Total (2)	£

10.2 Extension of time

Fig 10.1—cont.

	costs/week
3 INTEREST	

Interest payable on estimated capital expended up to the contract completion date, but from which no benefit is derived. Estimated expenditure taken as 80 per cent of contract sum and fees.

Contract sum	£
Architect's fees (90 per cent)*	£
Quantity surveyor's fees (90 per cent)**	£
Consultants' fees (90 per cent)**	£
Salary of clerk of works	£
(£/wk x contract period)	£

Interest charges at current rate of £...per cent: interest therefore

$$= \frac{80\% \text{ capital expended} \times \text{interest}}{52}$$

Total (3)	£

It is essential that all costs are additional: i.e. they would not be incurred if the contract were completed on the contract completion date. The headings given are examples only; every job is different.

**Professional fees are taken as 90 per cent of total because some professional work remains to be done after practical completion.*

Fig 10.1—*cont.*

4 INFLATION
   Current rate of inflation ... %/year
   Total (1) x ... per cent x contract
      period (in years)   £
   Total (2) x ... per cent x contract
      period (in years)   £
                                           ─────
                           Total (4)   £

5 TOTAL LIQUIDATED AND
   ASCERTAINED DAMAGES/WEEK
                           Total (1)   £
                           Total (2)   £
                           Total (3)   £
                           Total (4)   £
                                           ─────
                         Final total   £

10.2 Extension of time

Fig 10.2
Certificate of non-completion under clause 2.6

```
To the employer


CERTIFICATE OF NON-COMPLETION


In accordance with clause 2.6 of the above contract,
I/we hereby certify that the contractor [insert
name] has failed to complete the works by the date
for completion stated in the appendix [or] within
the extended time fixed by me/us under clause 2.3,
namely [insert date].

Dated this    day of            19

[Signed]                        Architect



Copy: Contractor
```

Claims

Fig 10.3
Cancellation of certificate of non-completion under clause 2.6 where time subsequently extended

```
To the employer


CANCELLATION OF CERTIFICATE OF NON-COMPLETION


In accordance with clause 2.6 of the above contract,
I/we hereby cancel my/our certificate of
non-completion dated  [insert date], having made a
further extension of time of  [insert number] weeks
for completion, under clause 2.3 of the contract.

Dated this    day of           19

[Signed]                   Architect


Copy: Contractor
```

10.2 Extension of time

2.3 is such that we think that it imposes a duty on the contractor to notify you of any delay (or likely delay) to progress and that it is not confined to notifying you of the events listed later.

No particular form of notice is prescribed, but if the notification does not give sufficient detail, you can require the contractor to provide you with such further information as is reasonably necessary to enable you to discharge your functions under the clause.

If you think that the completion of the works is being, or is likely to be, delayed by one or more of the events specified in clause 2.4, you must grant the contractor in writing a 'fair and reasonable' extension of time for completion. You must do this 'so soon as [you are] able to estimate the length of the delay'. In making this estimate, you should take into account the overall proviso to the clause. This proviso is in two parts: the contractor is required constantly to use his best endeavours to prevent (not to reduce) delay (eg, by re-programming); and he must do all that may be reasonably required to your satisfaction to proceed with the works. What steps he must take depends on the circumstances of the case, but you cannot require him to accelerate progress, and his obligation stops short of expending substantial sums of money.

There is no specific time limit, but you must grant a fair and reasonable extension of time for completion 'so soon as [you are] able to estimate the length of delay' beyond the currently fixed completion date. The various decided cases suggest that you must do so as soon as reasonably possible, although where there are multiple causes of delay, you may have no alternative but to leave the decision until a later stage. Fig 10.6 is a suggested letter for awarding an extension.

Indeed, because you have a 12-week period from practical completion in which to grant further extensions of time if the circumstances warrant it, it may be argued that parsimony is the order of the day—though that is not the view of contractors! Failure properly to grant an extension of time may result in the contract completion date becoming at large, so you must be careful, although you have some room to manoeuvre. If you are tardy, the contractor will probably press you for a decision.

The contractor's written notice is not in fact a precondition (or 'condition precedent' in lawyer's language) to the grant of an extension of time, because the clause goes on to empower you to grant an extension of time up to 12 weeks after the date of practical completion 'whether upon reviewing a previous decision or otherwise and whether or not the contractor has given notice'. You must, therefore, consider the contract as a whole and decide whether any extension of time is justified even if the contractor has failed to notify you. Unless you do this within the time limit laid down, you will jeopardize the employer's right to liquidated damages.

Claims

Fig 10.4
Contractor's duties in claiming an extension of time (clause 2.3)

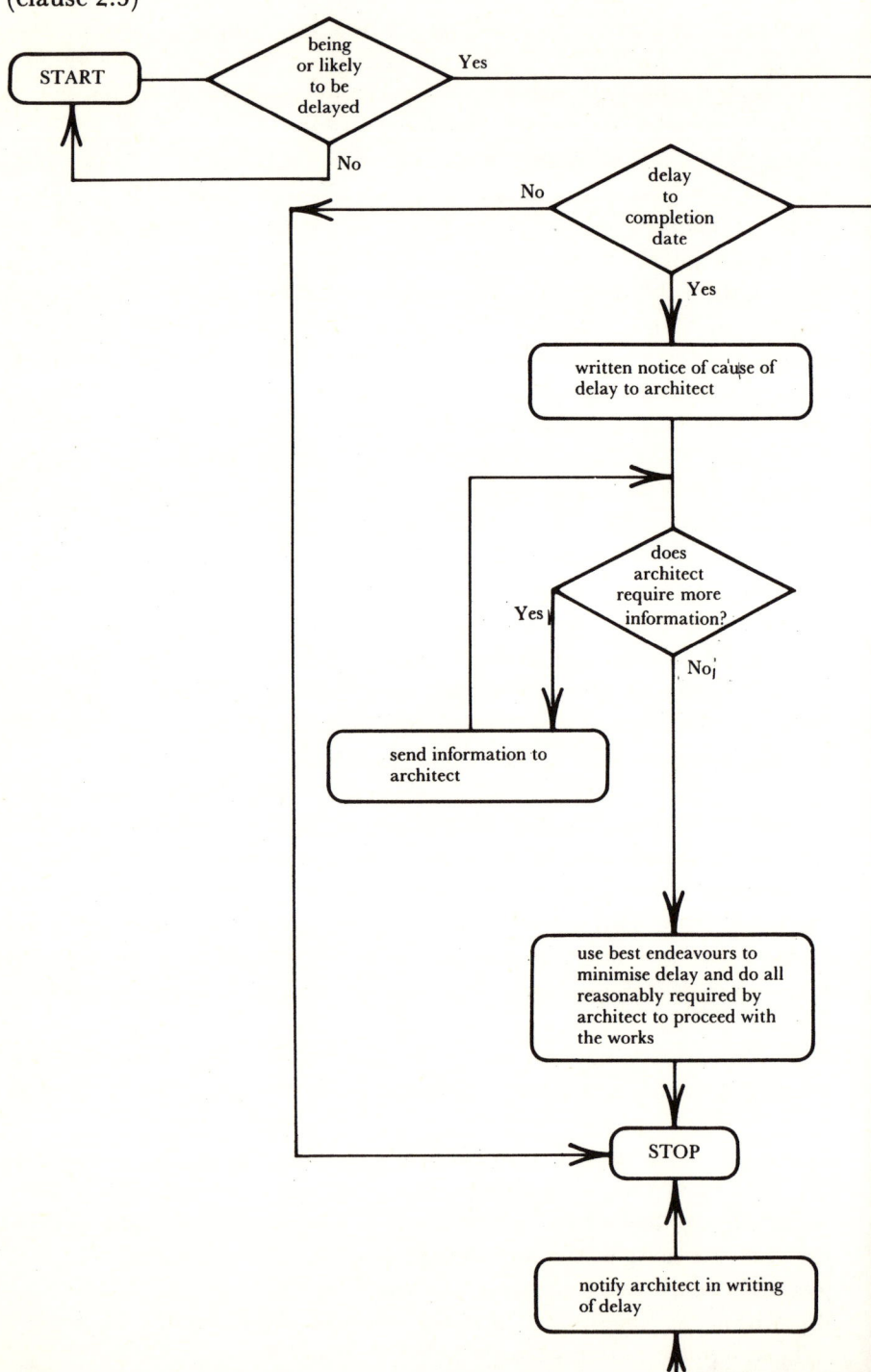

```
                    ┌──────────┐
                    │ being    │
    ┌───────┐      ╱ or likely ╲    Yes
    │ START │─────< to be       >────────────┐
    └───────┘      ╲ delayed   ╱              │
        ▲           └──────────┘              │
        │               │ No                  ▼
        │                              ┌──────────┐
        │              No             ╱ delay to  ╲
        ├─────────────────────────────< completion >
        │                              ╲ date     ╱
        │                               └──────────┘
        │                                   │ Yes
        │                                   ▼
        │                          ┌─────────────────────┐
        │                          │ written notice of   │
        │                          │ cause of delay to   │
        │                          │ architect           │
        │                          └─────────────────────┘
        │                                   │
        │                                   ▼
        │                              ┌──────────┐
        │                             ╱ does      ╲
        │                            ╱ architect   ╲
        │                    Yes    ╱ require more  ╲
        │                 ┌────────< information?    >
        │                 │         ╲               ╱
        │                 │          ╲       No    ╱
        │                 ▼           └──────────┘
        │        ┌──────────────────┐     │
        │        │ send information │     │
        │        │ to architect     │     │
        │        └──────────────────┘     │
        │                 │               │
        │                 └───────┬───────┘
        │                         ▼
        │              ┌────────────────────────┐
        │              │ use best endeavours to │
        │              │ minimise delay and do  │
        │              │ all reasonably required│
        │              │ by architect to proceed│
        │              │ with the works         │
        │              └────────────────────────┘
        │                         │
        │                         ▼
        │                     ┌──────┐
        └─────────────────────│ STOP │
                              └──────┘
                                  ▲
                                  │
                        ┌──────────────────┐
                        │ notify architect │
                        │ in writing of    │
                        │ delay            │
                        └──────────────────┘
```

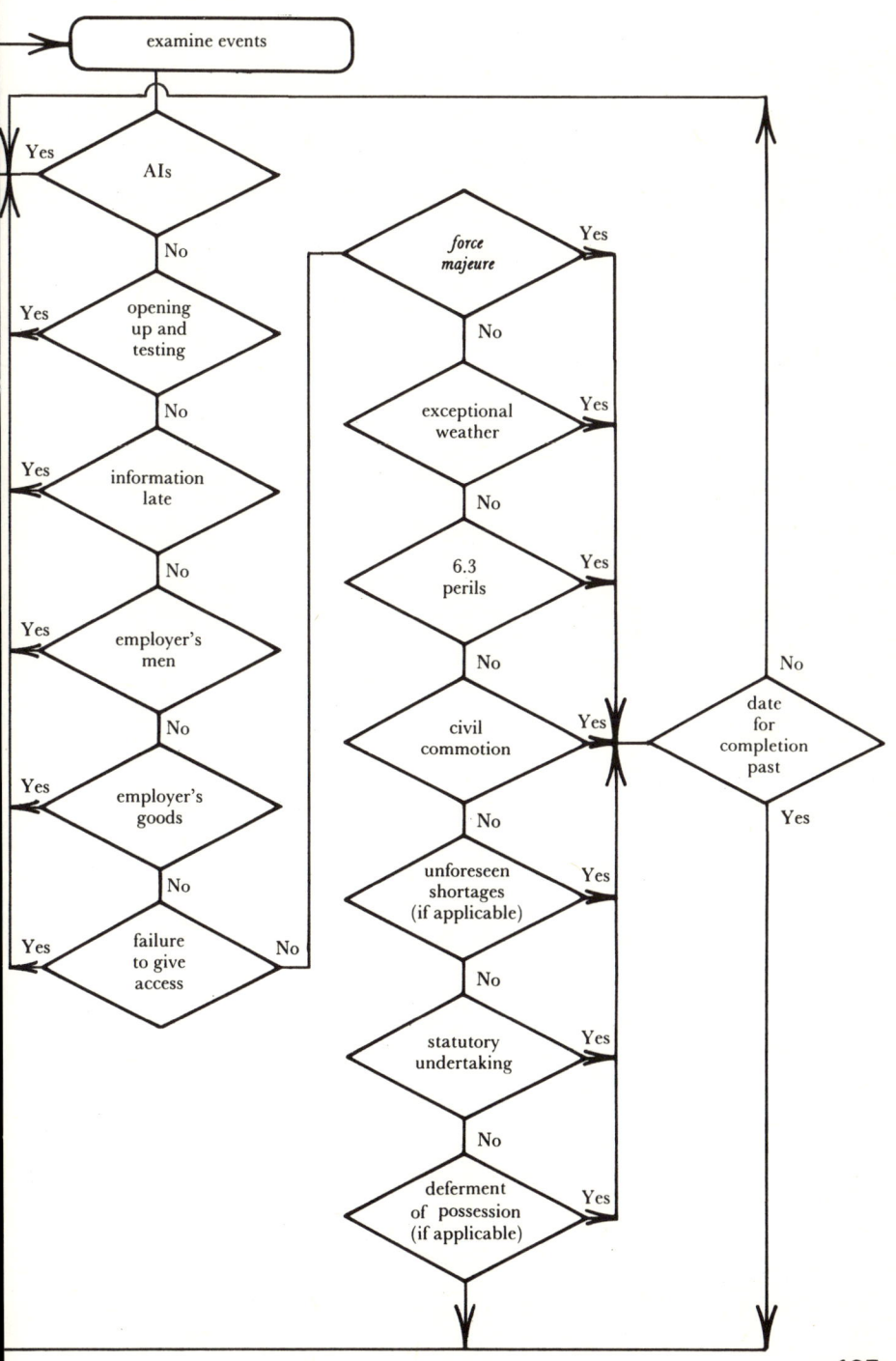

10.2 Extension of time

```
                    examine events
                          │
      ┌───────────────────┤
 Yes ←│ AIs │
      │ No
      │                       force majeure ── Yes →
 Yes ←│ opening up and testing │
      │ No                     │ No
      │                        │
 Yes ←│ information late │     exceptional weather ── Yes →
      │ No                     │ No
      │                        │
 Yes ←│ employer's men │       6.3 perils ── Yes →
      │ No                     │ No
      │                        │
 Yes ←│ employer's goods │     civil commotion ── Yes → ← date for completion past ── No →
      │ No                     │ No                        │ Yes
      │                        │
 Yes ←│ failure to give access │ No →  unforeseen shortages (if applicable) ── Yes →
                               │ No
                               │
                               statutory undertaking ── Yes →
                               │ No
                               │
                               deferment of possession (if applicable) ── Yes →
                               │
                               ▼
```

Claims

Fig 10.5
Architect's duties in relation to claim for extension of time (clause 2.3)

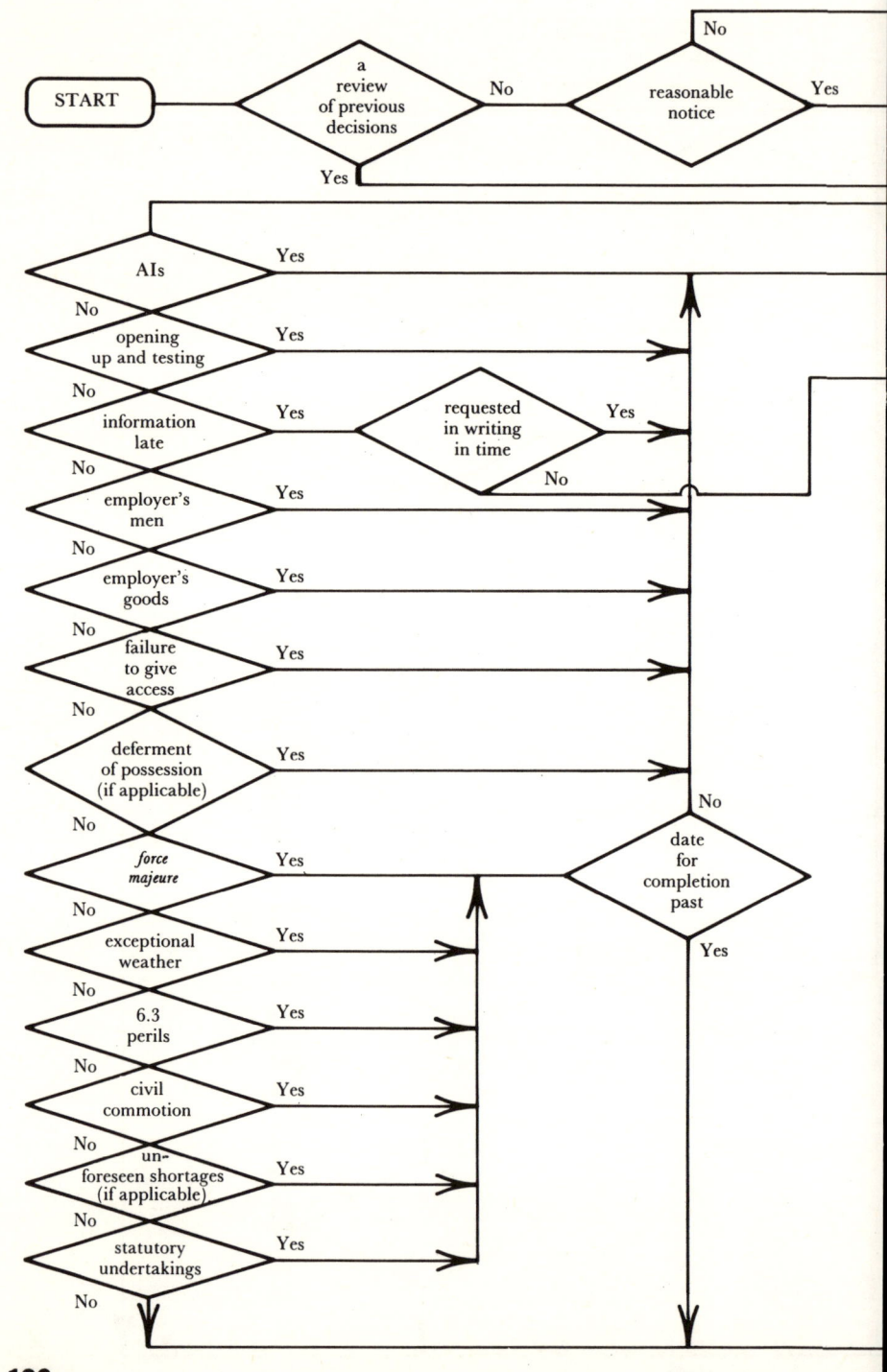

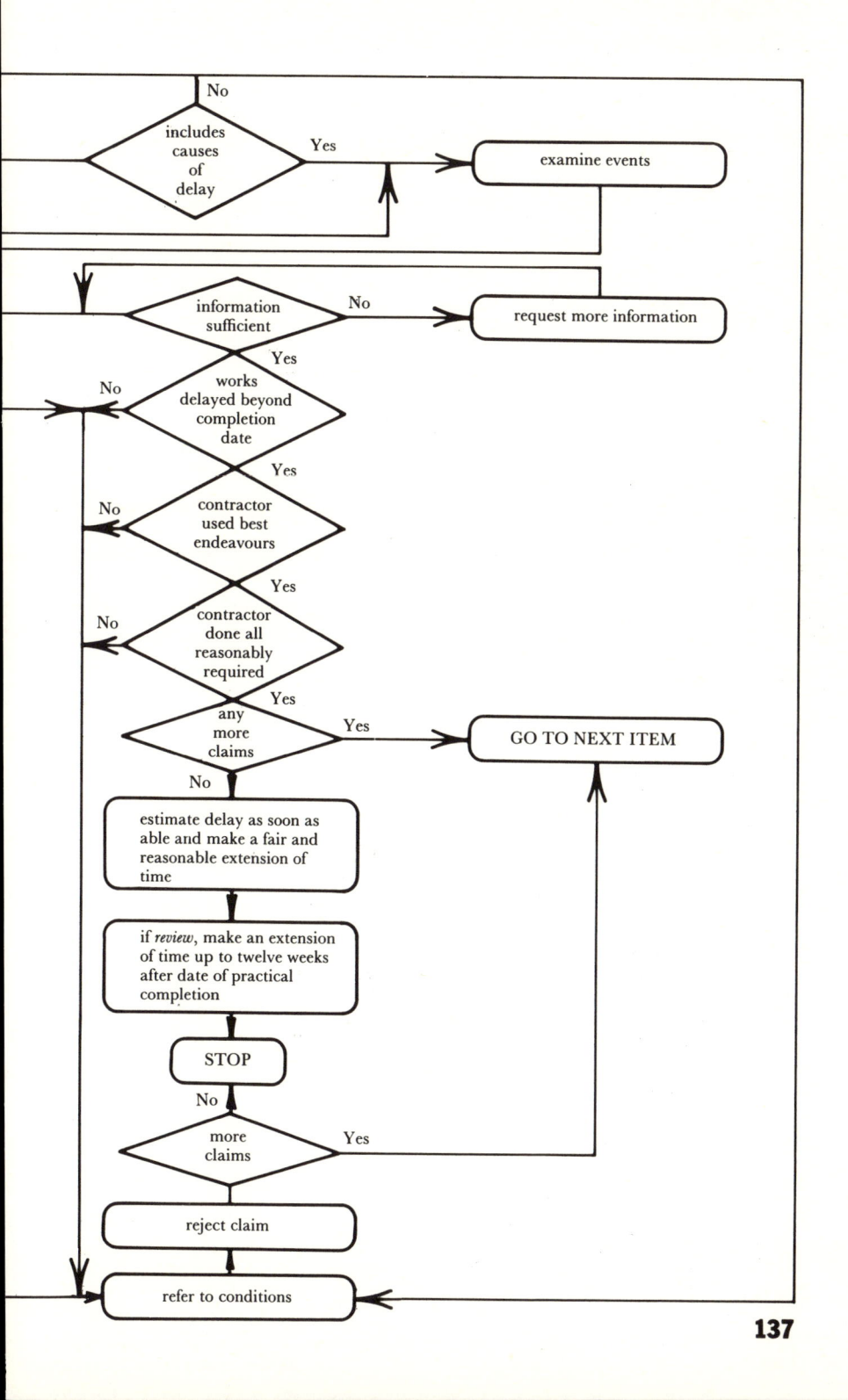

10.2 Extension of time

```
                    ┌─ No ─┐
                    │ includes │
                    │ causes   │─── Yes ───▶ examine events
                    │ of       │
                    │ delay    │
                    
                    │ information │─── No ───▶ request more information
                    │ sufficient  │
                         │ Yes
             ┌─ No ──│ works delayed beyond completion date │
                         │ Yes
             ┌─ No ──│ contractor used best endeavours │
                         │ Yes
             ┌─ No ──│ contractor done all reasonably required │
                         │ Yes
                    │ any more claims │─── Yes ───▶ GO TO NEXT ITEM
                         │ No
                    
                    estimate delay as soon as able and make a fair and reasonable extension of time
                    
                    if *review*, make an extension of time up to twelve weeks after date of practical completion
                    
                    STOP
                         │ No
                    │ more claims │─── Yes ───▶ (GO TO NEXT ITEM)
                         │
                    reject claim
                    
                    refer to conditions
```

Fig 10.6
Architect to contractor, granting extension of time under clause 2.3

Dear Sirs

I/we refer to your notice of delay of [*insert date*]. [*If appropriate add*] and the further information provided in your letter of [*insert date*].

In accordance with clause 2.3 of the contract, I/we hereby grant you an extension of time of [*specify period*], the revised date for completion now being [*insert date*].

Yours faithfully

10.2 Extension of time

The only effect of the contractor's failing to notify you as the clause requires is that he loses the benefit of an extension of time during the currency of the contract, but he is still entitled to it after practical completion if the circumstances warrant it. This is specially important if delay has been caused by one of the specified events which are the employer's responsibility in law (eg, late instructions or details). So we suggest that as a matter of practice you review every contract, and write to the contractor accordingly, even though the contract does not require you to do so.

Fig 10.7 is a suitable letter to send.

Note that when considering extensions of time after practical completion, you cannot reduce any extension of time previously granted.

IFC84 clarifies a moot point under other standard form contracts, since it expressly empowers the granting of an extension of time in respect of specified events which occur after the original or extended completion date has passed but before practical completion is achieved.

The events in question are the following:

— compliance by the contractor with your instructions about inconsistencies (clause 1.4), variations (clause 3.6), provisional sum expenditure (clause 3.8), postponement of work (clause 3.15), and named subcontractors (clause 3.3, as specified in that clause)

— compliance with your instructions requiring opening up for inspection or testing where the results are in the contractor's favour (clauses 3.12, 3.13.1)

— late instructions or drawings (eg, clause 1.7)

— execution of work not forming part of the contract by, or on behalf of, the employer, or failure to execute such work (clause 3.11)

— the employer's supply, or failure to supply, goods or materials which he has undertaken to supply

— failure by the employer to give any agreed access to or from the site.

Where such an event occurs after the completion date but before practical completion, you must grant a fair and reasonable extension of the contract period for any resultant delay. This provision clarifies the position considerably and, in practice, it prevents endless arguments with contractors, whose traditional view is that under other forms of contract you cannot, for example, order additional work once the completion date has passed. IFC84 makes it clear that you can do so, but you have a corresponding duty to extend time.

10.2.4 Grounds

Fourteen events which may give rise to an extension of time are specified in clause 2.4, and you can extend time only if one or more of

Fig 10.7
Architect to contractor after reviewing extensions of time

Dear Sirs

In accordance with clause 2.3 of the contract, I/we have now considered the contract progress as a whole and [*either*] I/we confirm the date of completion (as extended) as being [*insert date*] [*or*] I/we hereby grant you an extension of time of [*specify period*] to take account of [*specify*], thus giving a revised date for completion of [*insert date*].

Yours faithfully

10.2 Extension of time

these events occurs and causes, or is likely to cause, delay to progress. The events listed are traditional and they largely parallel those in JCT80, although there are differences.

The events divide into two distinct and separate groups: namely, those which merely entitle the contractor to an extension of time, and those which may also, quite independently, found a claim for loss and/or expense under clause 4.11.

Events giving rise to time only are as follows:

Force majeure

Although wider in its meaning than the English term 'Act of God', this has a restricted meaning here, because many matters which would otherwise constitute *force majeure* are dealt with expressly. For an event to amount to *force majeure*, it must be catastrophic and outside the control or contemplation of either contracting party (clause 2.4.1).

Exceptionally adverse weather conditions

The key word is 'exceptionally', since the contractor is expected to programme to take account of the sort of weather normally to be expected in the area at the relevant time of year. Moreover, the mere occurrence of exceptionally adverse weather—such as a fall of snow in the Midlands in mid-June—is not sufficient. It must actually delay progress (clause 2.4.2).

Clause 6.3 perils

These are the usual insurance risks—fire, tempest, and so on (clause 2.4.3).

Strikes and similar events

For the most part, the list of events is self-evident, and the breadth of the strike clause should be noted. It covers strikes, etc, affecting any of the trades employed on the works and also those engaged in preparing or transporting materials, etc, needed for the works (clause 2.4.4).

Shortage of labour and materials

These are optional grounds and will rank for extension of time only if stated in the appendix to apply. The limitation should be noted: the inability must be for reasons beyond the contractor's control; and the shortage, etc, must not have been reasonably foreseeable at the date of

141

Claims

tender—ie, the date 10 days before the date fixed for receipt of tenders by the employer (see clause 8.3). It is suggested that clauses 2.4.10 and 2.4.11 should be stated to apply in the appendix unless it is reasonable in all the circumstances to expect contractors tendering for the work to take the risk that labour or materials or both will not be available (clauses 2.4.10 and 2.4.11).

Local authority or statutory undertaker's work

This covers only work done 'in pursuance of . . . statutory obligations in relation to the works' or failure to carry out such work. If a statutory undertaker (eg, a gas board) does work under contract with the employer, this would fall under clause 3.11 and be dealt with under clause 2.4.8 as regards extension of time (clause 2.4.13).

The important point about all the foregoing events is that delay from one of them does not give rise to any claim for extra cost but merely to a claim for extension of time. Their common characteristic is that they are outside the control of either contracting party, whereas the second group consists of events which are the responsibility of the employer, either personally or through those for whom he is vicariously responsible in law.

Events giving rise to time and, independently, to direct loss and/or expense are as follows:

Architect's instructions

Those referred to are: clause 1.4, inconsistencies; clause 3.6, variations; clause 3.8, expenditure of provisional sums; clause 3.15, postponement of any work to be executed under the contract; clause 3.3, named subcontractors to the extent specified in clause 3.3 itself (clause 2.4.5); clauses 3.12 and 3.13.1, opening up or testing and failure of work—to the extent that the contractor is not at fault (clause 2.4.6).

Late instructions and drawings

The instructions must be those empowered by the contract (see Table 4.2), and there are limitations. The contractor must have made a specific application at the right time, and this is a matter where there is much room for argument. The wording is clear—the contractor must make a specific written application in each case, and what is a reasonable time does not depend solely on the contractor's convenience. He must have need of the information and be able to act on it, and the reasonableness of the timing must take account of your workload and commitments as well as his (*Neodox* v *Borough of Swinton &*

10.2 Extension of time

Pendlebury [1958]). The right time is to be judged 'having regard to the date for completion' as extended. It is not clear what the position is if the contractor does not make an application for information and you are in fact late in supplying it. The content assumes that you will supply information as the works progress and failure to do so is probably a breach of contract for which the employer is responsible (see clause 1.7). It does not appear to be a matter which you can take into account when considering the question of extensions of time within the 12-week period from practical completion (clause 2.4.7).

Work not forming part of the contract

Clause 3.11 enables the employer to carry out such work either himself or through others, and, because of the breadth of the wording, this ground extends to cover a situation where, for example, such work is done badly by direct contractors and delay is caused by subsequent remedial work (clause 2.4.8).

Supply or non-supply of materials by employer

This applies only where the employer has so agreed (clause 2.4.9).

Employer's failure to give agreed access

This is not so extensive as might at first sight appear. The wording should be noted. An extension of time cannot be granted on this ground if, for example, the employer fails to obtain a wayleave over a third party's property and certainly will not extend to deal with the situation where site access is impeded by third parties for whom the employer has no responsibility in law (clause 2.4.12).

Deferment of possession

This will give rise to an extension of time when clause 2.2 is stated to apply and an extension of time is an almost necessary concomitant of deferment of possession.

10.3 Loss and expense claims

10.3.1 Definition

Clause 4.11 gives the contractor a right to be reimbursed 'direct loss and/or expense' which he incurs as the result of the events specified and for which there is no other payment under the contract, provided he follows the procedure laid down. Claims for loss and expenses are a regulated provision for the payment of damages.

'Direct loss and/or expense' is to be equated with the damages recoverable for breach of contract at common law. The purpose of the clause is to set out the contractor's rights in anticipation of the specified events and is the means of putting the contractor back into the position in which he would have been but for the delay or disruption.

Since the settlement amounts to damages at common law, an exact establishment of the contractor's additional costs must be made. Many contractors hold a fallacious view about claims generally, and the following three important points are often overlooked:

— the contractor must 'mitigate his loss' ie, take reasonable steps to diminish his loss, for example by redeploying his resources

— since damages are subject to the common-law 'foreseeability test', the contractor can recover only that part of the resultant loss and/or expense that was reasonably foreseeable to result. This is to be judged at the time the contract was made and not in the light of the events which have occurred

— the loss and/or expense must be direct. It must have been caused by the event relied on without any intervening cause.

10.3.2 Procedure

The contract requires the contractor to make a written application to you within a reasonable time of its becoming apparent that he has incurred, or is likely to incur, loss or expense resulting from specified causes. These are that the employer defers giving possession of the site under clause 2.2 (if applicable); *or* that the regular progress of the works is 'materially' affected by one or more of the seven matters set out in clause 4.12. 'Materially' means to a substantial extent.

The contractor's written application need not be in any particular set form, but you may require him to provide further information to enable you (or the quantity surveyor) to assess the claim. The sort of information you may require will depend on the circumstances, and it is the contractor's duty to provide you (or the quantity surveyor) with sufficient documentary evidence to enable the claim to be dealt with.

When the contractor has made a written application, you must decide whether he has incurred, or is likely to incur, loss and/or expense resulting from one or more of the specified matters, and you must also be satisfied that the contractor is not receiving payment in respect of it under some other contract provision (eg, under clause 3.7 in respect of variations).

If you are so satisfied, then you must ascertain the amount of the loss or

10.3 Loss and expense claims

expense, or instruct the quantity surveyor to do so on your behalf. Figures cannot be plucked out of the air: 'ascertain' means to establish definitely and is not the same as 'estimate' or 'guess'. The amount so ascertained is to be included in the next payment certificate.

The contract allows additional or alternative claims for breach of contract based on the same facts, as is made clear by the last sentence of clause 4.11, which preserves the contractor's normal rights at commonlaw. Because of this provision, the contractor may also bring commonlaw claims, which are often based on implied terms relating to non-interference with his progress. You have no power or authority to deal with commonlaw claims: the only claims which you can deal with are those arising under clause 4.11. From the contractor's point of view, the fact that his commonlaw rights are preserved is a very real benefit, because if he has failed to make the written application required by clause 4.11, and so has lost his right to reimbursement under the contract terms, he can bring a claim at common law based on the same facts, given that the event relied on is also a breach of contract (eg, late instructions from you). Such claims, however, must be pursued in arbitration or litigation.

The flowchart Fig 10.8 sets out the contractor's duties under clause 4.11. Flowchart Fig 10.9 sets out your duties under the provision.

10.3.3 Matters grounding a claim

Seven 'matters' are listed in clause 4.12, and it is the occurrence of one of these—or the employer's deferment of giving possession of the site—which triggers off a claim.

The clause 4.12 matters are as follows:

—   late instructions: the paragraph refers to the contractor's not having received necessary instructions, etc, from you in due time. Two conditions must be met before you can allow a claim under this head: the contractor must have specifically applied to you in writing for the information; and his application must have been made in due time

—   opening for inspection: the inspection or test must have shown that the work, materials, or goods were in accordance with the contract if a claim is to be made

—   execution of work by those engaged by the employer: see clause 3.11

—   supply or non-supply of materials by employer: the employer must have agreed with the contractor to so supply

—   postponement of work: see clause 3.15

—   failure to give ingress or egress: see clause 4.12.6

—   architect's instructions: (clause 4.12.7) those referred to are issued under clauses 1.4 (inconsistencies), 3.6 (variations), 3.8 (provisional sums), and 3.3 (named subcontractors: as specified in that clause).

Claims

Fig 10.8
Contractor's duties in claiming loss and/or expense (clause 4.11)

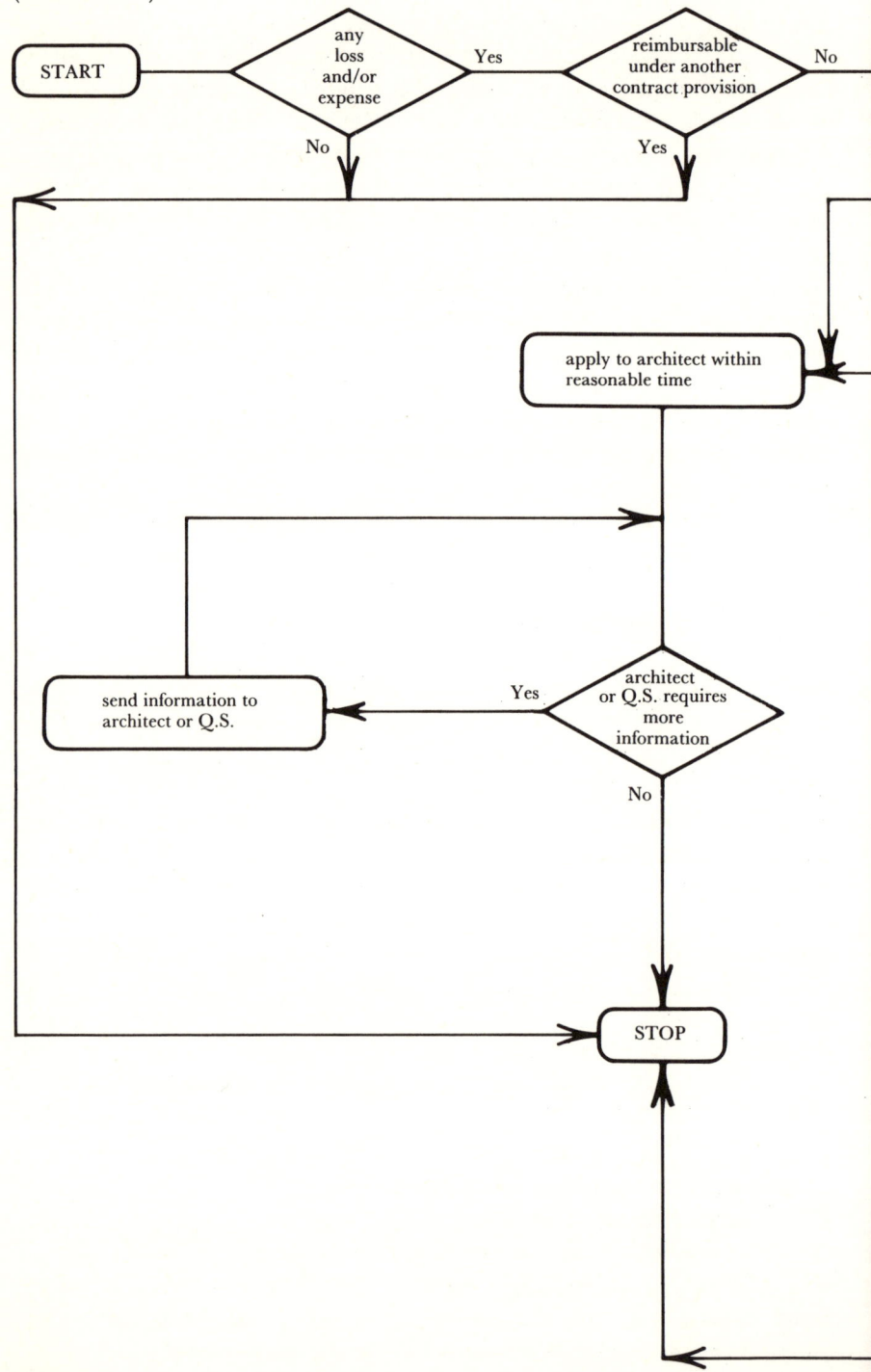

```
START → any loss and/or expense?
  Yes → reimbursable under another contract provision?
    No → apply to architect within reasonable time → architect or Q.S. requires more information?
      Yes → send information to architect or Q.S. → (loop back)
      No → STOP
    Yes → STOP
  No → STOP
```

10.3 Loss and expense claims

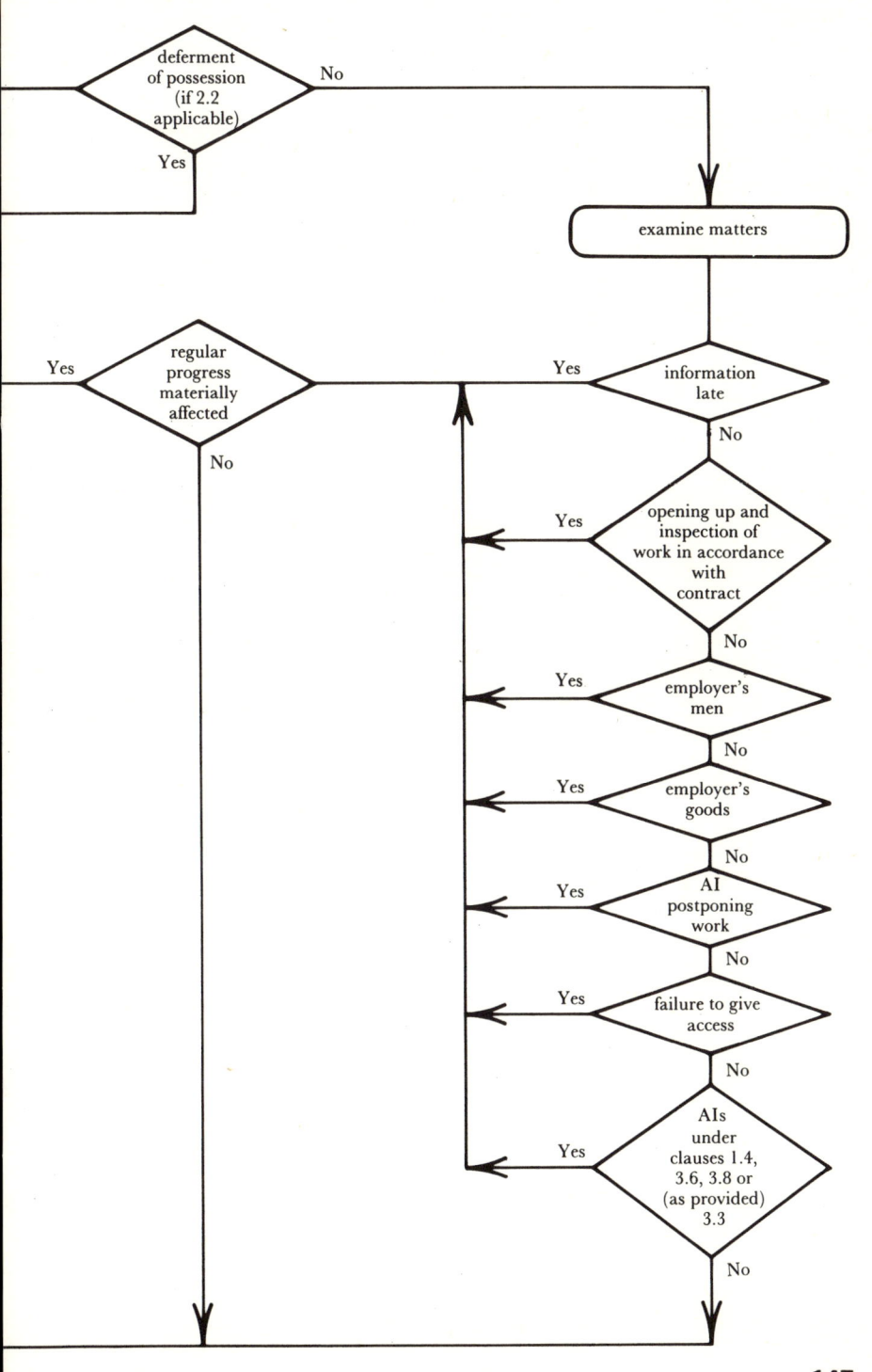

```
                    deferment
                  of possession         No
                    (if 2.2 applicable) ────────────┐
                    Yes                             │
                     │                              ▼
                                              examine matters
                                                    │
                                                    ▼
  Yes    regular progress            Yes       information
  ◄──── materially affected  ◄──────────────── late
         │                    │                     │ No
         │ No                 │                     ▼
         │                    │             opening up and
         │                    │     Yes     inspection of
         │                    ◄──────────── work in accordance
         │                    │             with contract
         │                    │                     │ No
         │                    │     Yes      employer's
         │                    ◄──────────── men
         │                    │                     │ No
         │                    │     Yes      employer's
         │                    ◄──────────── goods
         │                    │                     │ No
         │                    │     Yes       AI postponing
         │                    ◄──────────── work
         │                    │                     │ No
         │                    │     Yes      failure to give
         │                    ◄──────────── access
         │                    │                     │ No
         │                    │                 AIs under
         │                    │     Yes       clauses 1.4,
         │                    ◄──────────── 3.6, 3.8 or
         │                                  (as provided) 3.3
         │                                         │ No
         ▼                                         ▼
```

Claims

Fig 10.9
Architect's duties in relation to a claim for loss and/or expense (clause 4.11)

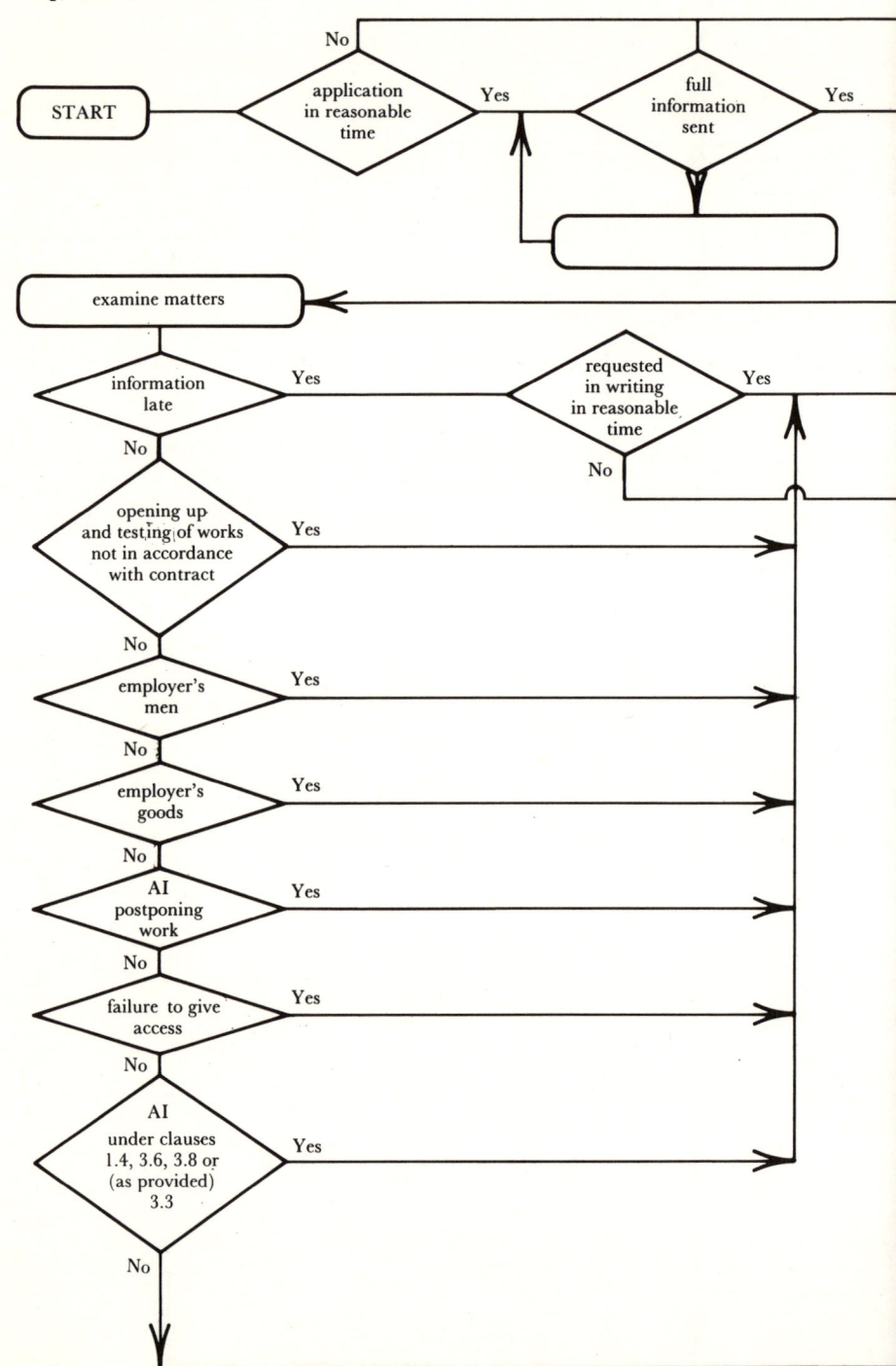

START → application in reasonable time?
- No →
- Yes → full information sent?
  - No → (loop back)
  - Yes → examine matters

examine matters →
- information late? Yes → requested in writing in reasonable time?
  - Yes →
  - No →
- No ↓
- opening up and testing of works not in accordance with contract? Yes →
- No ↓
- employer's men? Yes →
- No ↓
- employer's goods? Yes →
- No ↓
- AI postponing work? Yes →
- No ↓
- failure to give access? Yes →
- No ↓
- AI under clauses 1.4, 3.6, 3.8 or (as provided) 3.3? Yes →
- No ↓

10.3 Loss and expense claims

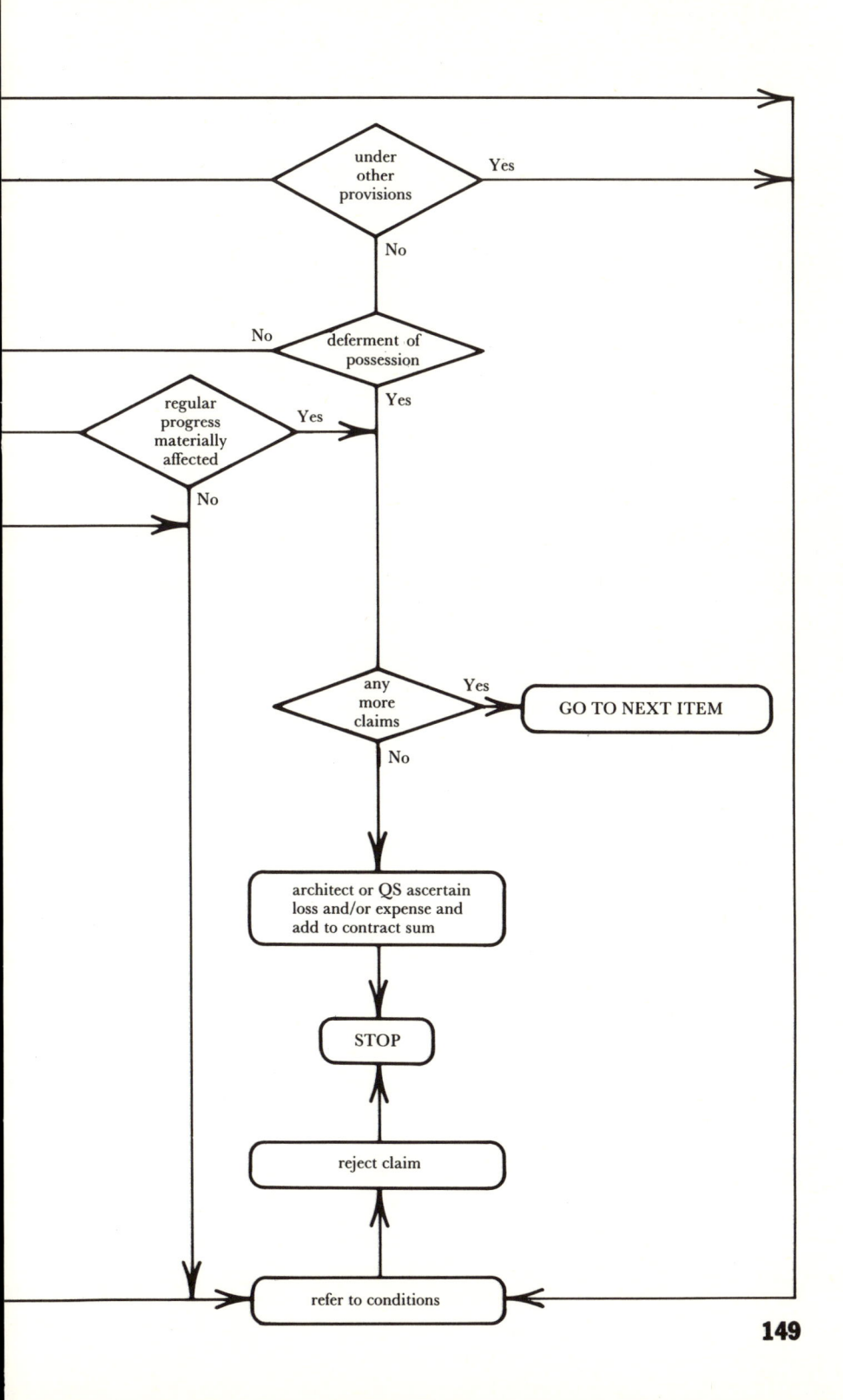

- under other provisions — Yes →
- No ↓
- deferment of possession — No →
- Yes ↓
- regular progress materially affected — Yes →
- No →
- any more claims — Yes → GO TO NEXT ITEM
- No ↓
- architect or QS ascertain loss and/or expense and add to contract sum
- STOP
- reject claim
- refer to conditions

149

Claims

10.4 Summary

Claims for time and money are distinct; there is no necessary connection between the two.

Liquidated damages

Liquidated damages are:
— a genuine pre-estimate of likely loss or a lesser sum
— recoverable without proof of loss
— recoverable by deduction under the contract only if you have issued a certificate of delay *and* the employer has notified the contractor in writing of his intention to require liquidated damages.

Extension of time

You are bound to grant a fair and reasonable extension of time for completion on the happening of certain events
— failure to do so may result in the contract date becoming 'at large' and liquidated damages being irrecoverable.
The contractor must observe the contract notice procedure and, if he does so, you must:
— grant in writing a fair and reasonable extension of time for completion as soon as practicable
— review contract progress within 12 weeks of practical completion and adjust the contract time accordingly, even if the contractor has not notified you of an event giving rise to extension of time
— grant an extension of time only on grounds specified in the contract.

Loss and expense claims

The contractor has a right to be reimbursed 'direct loss and/or expense' under the contract on the happening of certain disruptive events provided he invokes the contract procedures.
The contractor must:
— make written application to you at the right time
— provide you with the necessary supporting evidence.
The architect must:
— be satisfied that the notified event is a valid ground of claim
— ascertain (or instruct the quantity surveyor so to do) the amount of loss or expense directly incurred

10.4 Summary

— include the sum so ascertained in the next certificate of payment.

The architect has no power under the contract to deal with common law or *ex gratia* claims.

11 Payment

11.1 The contract sum

The figure shown in article 2 of the agreement is of great importance. It is the amount for which the contractor has agreed to carry out the whole of the work. In contracts such as IFC84, which are known as 'lump-sum contracts', the contractor is entitled to payment provided he substantially completes the whole of the work. The fact that the contract provides the interim payments does not alter the position. If he abandons the work before it is finished, the employer is entitled to pay nothing more. The principle of interim payments is to provide sufficient money to allow the contractor to carry out the work. It is purely a business arrangement. It matters not that the contractor has imparted considerable benefit to the employer. If he does not substantially complete the work, he is not entitled to payment (*Hoenig* v *Isaacs* [1952]). In practice, this severe view is somewhat modified if the employer determines the contractor's employment (see section 12.1.7).

The contract sum may be adjusted only in accordance with the provisions of the contract (see Table 11.1). Errors or omissions of any kind in the computation of the contract sum are deemed to have been accepted by employer and contractor (clause 4.1). The only exceptions to that are those instances specifically provided for (see section 3.1.3). They are inconsistencies in or between documents; errors or omissions in description or quantity; errors or omissions in the particulars relating to named persons; and departures from the Standard Method of Measurement. The contractor may make all kinds of errors in pricing the contract documents. He may under- or over-price items, overlook

11.1 The contract sum

Table 11.1
Contractual provisions which allow adjustment of the contract sum

Clause	Provision
2.10	Deduction for defects not to be made good
3.3.4(a)	Adjustment after determination of named person's employment
3.7	Additions or deductions in respect of instructions requiring a variation
3.9	Deduction for setting out errors not to be amended
3.12	Additions to cover the cost of opening up or testing if the work is in accordance with the contract
4.5	Final adjustment of the contract sum
4.9	Adjustment to take account of fluctuations
4.10	Adjustment to take account of fluctuations in respect of named persons
4.11	Additions for loss and/or expense
5.1	Additions for statutory fees and charges
6.2.4	Additions for insurance for the liability, etc, of the employer
6.3B.2	Additions for employer's failure to insure
6.3C.1	Additions for employer's failure to insure

items, or simply make a mistake in adding up totals. Once his price is accepted, however, it may not be altered. The quantity surveyor will have checked the contractor's calculations before the tender was accepted, but he may not have noticed the error. It is always bad, from the point of view of both contractor and employer, if the contractor finds, after the contract is in being, that he has made an error which will result in his losing money. He will often attempt to recoup his losses by submitting claims at any opportunity. So particular care must be exercised, before recommending a tender for acceptance, that your checks or the quantity surveyor's have been through and that the final figure does not appear suspiciously low in comparison with other tenders.

11.2 Payment before practical completion

11.2.1 Method and timing

The parties may make whatever arrangements they wish for interim payments. Where the contract is of relatively high value, it is customary to pay at monthly intervals, but if the value is low or the priced documents make it convenient, it may suit both parties to agree that payment will be made on the completion of certain defined stages. The parties may have the agreement set out in the specification, schedules of work, or bills of quanties before entering into the contract, or they may agree the mode of payment before commencing work. If a particular system of payment is desired, it is best to have it set out in the contract documents at tender stage because
— the method and regularity of payment will significantly influence the contractor's tender, and because,
— if no agreement to the contrary is concluded after the parties have entered into a contract, the provisions of clause 4.2 will apply.
The contractual provisions are straightforward, but certain points will repay careful study.
Payment is to be made after you issue your certificate. The interval between certificates is to be one month (ie, one calendar month) unless a different interval is stated in the appendix. The intervals are calculated from the date of possession. In the case of deferment, it is suggested that, to avoid absurdity, the intervals are to be calculated from the new or deferred date of possession. Your first certificate is due one month from the date of possession. It must include the total of the amounts as specified in clauses 4.2.1 and 4.2.2 at a date not more than seven days before the date of the certificate. The employer has 14 days from the date of the certificate to make payment.
If a calendar month is some four weeks, the contractor could wait six weeks from the date of possession before receiving his first payment for three weeks' work. The importance of prompt payment cannot be over-emphasised. It is a means of assisting the contractor's cash flow, reducing his overdraft requirements and hence the interest he has to pay, and so increasing his chance of making a reasonable profit, thus making him less likely to resort to the submission of claims.

11.2.2 Valuation

Whenever you feel that it is necessary to do so, you are empowered to request the quantity surveyor to carry out a valuation before you issue your certificate. The clause says 'necessary for the purpose of

11.2 Payment before practical completion

ascertaining the amount to be stated as due'. To ascertain is to find out for certain, so you will take this step when you feel unable to know for certain without the aid of the quantity surveyor. Because the process of valuation is specialised, there will be few instances, in practice, when you will not require the quantity surveyor's assistance if you are wise.

The amount to be included on your certificate is to be the total of amounts in clauses 4.2.1 and 4.2.2, less only any amounts included on previous certificates—note, not any amounts previously paid, but any amounts previously certified. You need not concern yourself, when certifying, whether the employer has paid or paid in full. The amounts you must include are divided into two categories: those items of which you are to include only 95 per cent of their value (ie, the employer will keep 5 per cent as retention); and those whose full value must be included (ie, no retention).

The items are discussed in section 11.2.3. The question of value requires more consideration. There is a very large difference of opinion about the meaning of the word 'value' in this context. It is of great importance because, in the event of the contractor's insolvency, it is essential that you have not overcertified. Certification is your responsibility whether or not you have worked from the quantity surveyor's valuation. If you disagree with his valuation, your duty is to change it. You must ensure that the quantity surveyor puts no value on defective work. He cannot be expected to know what you consider to be defective unless you tell him. Do so in writing, every month before he carries out his valuation.

So what is the value of the contractor's work? One school of thought, generally accepted throughout the industry, is that the value of the contractor's work is to be ascertained by reference to the priced contract document. His entitlement is to be payment for the work he has properly done at the rates he has inserted in the document. It seems reasonable. Defective work is not included, and a sum of money is retained against problems arising. The biggest problem which could arise is that the contractor goes into liquidation immediately following a payment.

The second school of thought derives from this pessimistic outlook. The value of the contractor's work, from the employer's point of view, is the value of the whole contract less the cost of completing with the aid of another contractor and with additional professional fees. The additional cost to the employer in such circumstances is considerable. The retention fund, even at a late stage, would not cover it.

In operating the latter system, certificates in the early stages of a contract might be very low or even for minus figures. The chief difficulty is in deciding how much it would cost to complete the contract at the time of each valuation. (Here, the quantity surveyor, in his role as building economist, bears a heavy responsibility). There are, therefore, two possible interpretations to put on the word 'value': the value to the builder, or the value to the employer. It is for you to decide which one you will adopt. The second system may seem harsh to the contractor, but it has the merit of ensuring an adequate supply of funds if the contractor has to abandon the work. Contractors obviously do not like this approach, and it is fair to inform them that you are adopting the system at the time of tender. Higher tenders will result, but you will not have to account to the employer for the additional costs of completion if the contractor fails. Moreover, recent case law suggests that the first view is the better one.

11.2.3 Amounts included

It is logical to consider the amounts separately in the two categories. Amounts on which the employer is allowed a five per cent retention are the following:

— the total value of the work properly executed by the contractor, including items classed as variations (clause 3.7), and formulae adjustment (clause 4.9(b)). The interpretation you are to put on 'value' has already been discussed (section 11.2.2).

— the total value of materials which have been reasonably and not prematurely delivered to, or adjacent to, the works for incorporation, provided that they are adequately protected against weather and damage. You need not include in your certificate any materials which the contractor has clearly delivered to site for the express purpose of obtaining payment. Clause 1.10, unfixed materials, and clause 3.2.2, subcontracting, are intended to ensure that materials paid for in this way by the employer become his property in law (clause 4.2.1(b)).

— the value of any off-site materials, at your discretion. The inclusion of off-site materials in your certificate has the potential to raise serious problems. Clause 1.11 is intended to ensure that, once the employer has paid for them, the materials become his property. The contractor is not permitted to remove the materials, or to allow anyone else to do so, except for use on the works. All the while, the contractor is to remain responsible to the employer for any loss or damage. There are two dangers. First, that the supplier may have incorporated a retention-of-title clause in the contract of sale to the contractor. That means that, despite anything which might be written into this contract, the materials remain the property of the supplier until he receives payment from the contractor. Compare these provisions with the provisions for subcontractor's materials in clause 3.2.2 (see section 8.2.1). You would be prudent to insist on inspecting the supplier's conditions of sale before you consider including the value of off-site materials. Second, it

11.2 Payment before practical completion

is difficult to be sure that the materials you inspect at the contractor's yard are not really intended for some other job. Unlike JCT80 clause 30.3, IFC84 contains no provisions which adequately protect the employer's interest. It is suggested that you must adopt JCT80 clause 30.3 provisions. Even then, the situation is not absolutely safe. It has been known for a contractor to label and set aside, say, sink units for a particular contract until after inspection by the architect, then relabel them for the benefit of another architect and a different contract. That is sharp practice, but, if the contractor goes into liquidation, there is no consolation for the employer. The rule must be that you should not include off-site materials unless you decide that it would be quite unfair not to do so, and you have satisfied yourself, as far as it is possible to do so, that the employer's interests are fully protected. Remember, if something goes wrong, the employer will be quick to inform you that he paid only on your advice. You might be liable.

Amounts on which the employer is allowed no retention are the following:

— in accordance with clause 3.12, where the contractor has carried out opening up and/or testing and the work is found to be in accordance with the contract. The contractor is entitled to be paid the cost of the work and the cost of making good

— in accordance with clause 4.9(a), contribution, levy, and tax fluctuations

— in accordance with clause 4.10, named-person fluctuations

— in accordance with clause 4.11, where the contractor is entitled to payment of direct loss and/or expense due to disturbance of regular progress

— in accordance with clause 5.1, where the contractor has paid statutory fees or charges which were not provided for in the contract documents

— in accordance with clause 6.2.4, where a provisional sum is included to cover special insurance which is the liability of the employer

— in accordance with clause 6.3A.4, where the contractor has insured the works and an insurance claim has been accepted

— in accordance with clause 6.3B.2, where the employer, not being a local authority, has failed to insure and the contractor can produce evidence that he has taken out the appropriate insurance himself

— in accordance with clause 6.3C.1, where the employer has failed to insure and the contractor can produce evidence that he has taken out the appropriate insurance himself.

The amounts payable will depend on the amounts ascertained at the time of the valuation. Deductions are to be made in this category as follows:

Payment

— in accordance with clause 3.9, where you have instructed that errors in setting out should not be amended and an appropriate deduction should be made

— in accordance with clause 4.9(a), contribution, levy, and tax fluctuations as appropriate

— in accordance with clause 4.10, named-person fluctuations as appropriate.

Not included in this clause, probably by oversight, is the provision in clause 2.10 that if you instruct that defects, shrinkages, and other faults are not to be made good, an appropriate deduction is to be made.

11.3 Payment at practical completion

Although it is nowhere expressly stated, regular interim certificates and payments will cease at practical completion, simply because there will be no further work to certify. It may be that a particularly difficult claim will not be settled until a month or two after practical completion. You should make certain that, in such a case, money is released to the contractor whenever you are confident that any part of the claim is valid. There is no contractual liability upon you to do so, however, and you may, if you wish, wait until the whole claim has been quantified before certifying. You may not wait until the issue of the final certificate. It is probable that, if practical completion has passed by more than 14 days, the contractor can invoke clause 4.2 to compel you to issue a certificate once the claim has been ascertained.

The contract provides for a special payment to be made at practical completion (clause 4.3). You must issue a certificate within 14 days of the date of practical completion. The employer must pay, as before, within 14 days of the date of the certificate. The amounts to be included fall into the same categories (see section 11.2.3) as for interim certificates. There is an important difference, however. You must include 97½ per cent of the value of amounts in the first category, compared with 95 per cent included in that category before practical completion. This has the effect that the employer releases half the retention he has been holding. The remaining retention is held by the employer until the final certificate.

11.4 Retention

Retention is dealt with specifically by clause 4.3. The object of the clause is to safeguard the contractor's interest in the retention, but there are other implications. So it is curious that the initial provision excludes local authorities from its operation. This must be because the draftsman, while following accepted practice in assuming that a local

11.4 Retention

authority will not become insolvent, overlooked the clause's broader application.

The employer is stated to be a trustee and his interest in the retention to be fiduciary (which amounts to much the same). He is a trustee for the contractor. That means that the contractor has the right to insist that the retention fund be kept in a separate bank account clearly designated as held in trust for the contractor. This safeguards the money if the employer becomes insolvent. Some contracts now have a special provision requiring the separate account.

Although the employer holds the retention in trust for the contractor, the clause states that he has no obligation to invest—that is, no obligation to make the best use of the money on behalf of the contractor and to return him interest on it. Since this provision is contrary to statute, it is probable that it is a position the employer could not defend if a contractor decided to take action on it. However, the point has yet to be decided by the courts. Try not to start a test case!

The contractor's interest in the retention is said to be subject only to the employer's right to take money from it from time to time to pay amounts which the contract provides for him to deduct from sums due or to become due to the contractor. So this clause, which is not applicable to local authorities, is the only express term allowing the employer to use retention monies for other than the contractor's purposes. No doubt, in the complete absence of this clause, a term would be implied to allow the employer to deduct from the retention to make good the contractor's defaults. In the case of a local authority, the clause is not absent, but it is stated to apply only 'where the employer is not a local authority'. We suggest that local authorities should amend this clause with the aid of their expert advisors.

A list of the contract provisions referred to in this clause is given in Table 11.2.

Table 11.2
Contract provisions which entitle the employer to deduct from any sum due or to become due to the contractor

Clause	Provisions
2.7	Liquidated damages
3.5.1	Costs of employing other persons to carry out instructions
6.2.3	Contractor's failure to insure
6.3A.2	Contractor's failure to insure

11.5 Final payment

The contract lays down a particular time sequence for the events leading up to, and the issue of, the final certificate (see Fig 11.1).

The contractor has a duty to send to you, or to the quantity surveyor if you so instruct, all the documents which are reasonably required for the final adjustment of the contract sum (clause 4.5). He must send them either before practical completion or, more likely, within a reasonable time afterwards. In the context of this contract, it is probable that anything over one month would be unreasonable. You are entitled to request the kind of documents you require or to request particular documents, provided that your requests are reasonable in the circumstances.

Armed with this information, the quantity surveyor must prepare a statement of all the final valuations under clause 3.7. A copy of all the computations to arrive at the finally adjusted contract sum must be sent to the contractor before the end of the period of final measurement and valuation which is stated in the appendix. This period is usually—and if you insert no figure will be—six months. So if the contractor is unreasonably late in sending his documents, he cannot expect you to adhere to this timetable.

The contract does not state that the contractor must agree the finally adjusted contract sum before you issue your final certificate. In practice, it is customary to try to obtain agreement, and the contractor is usually sent two copies of the computations, one for him to sign as agreed and return. That is, no doubt, why the contract allows you 28 days from sending the computations or issuing your certificate of making good defects under clause 2.10, whichever is the later, to issue your final certificate (clause 4.6). The contract unaccountably allows 21 days for the employer or the contractor, as the case may be, to pay.

Your certificate must include the following:

— the total value referred to in clause 4.2.1(a)—ie, release of all the retention

— the amounts referred to in clause 4.2.2 finally ascertained, less any amount which is to be deducted under clause 2.10, defects liability; clause 3.9, levels; clause 4.9(a), contribution, levy, and tax fluctuations; or clause 4.10, named-person fluctuations

— less any sums previously certified.

If the result is in favour of the contractor, the employer will be liable to pay; if the result is in favour of the employer, the contractor will be liable to pay. The latter is a somewhat unusual situation.

11.6 The effect of certificates

Clause 4.8 provides that no certificate, other than the final certificate, is

11.6 The effect of certificates

conclusive evidence that any work, materials, or goods to which it relates are in accordance with the contract. All certificates, except the final certificate, are included, whether financial or not. For example, the issue of the certificate of practical completion does not prevent you from requiring the contractor to make good work not in accordance with the contract. The issue of an interim certificate is not evidence that all the work included is in accordance with the contract, otherwise you would not be entitled to omit defective work from one certificate if it had been included on a previous certificate. You undoubtedly have that power.

Even the final certificate is not conclusive (clause 4.7) as was once understood. There appears to be no good reason why it should be. The final certificate does have a conclusive effect in two respects:

—   where, and to the extent that, approval of the quality of materials or of the standards of workmanship is a matter for your opinion, the final certificate is conclusive that the quality and standards are to your reasonable satisfaction. The important nature of this clause, in conjunction with clause 1.1, has been discussed fully in section 4.2.1

—   the final certificate is conclusive evidence that the terms of the contract which require additions, deductions, or adjustments to the contract sum have been correctly operated.

There are two exceptions:

—   if there have been any accidental inclusions or exclusions of any items, or any arithmetical errors in any computation, they may be corrected

—   if any matter is the subject of proceedings commenced before, or within 21 days after, the issue of the final certificate, the certificate is not conclusive regarding that matter. Thus, either party has 21 days after the date of issue to refer to arbitration or to commence legal action through the courts. If the matter is not one about which the certificate is stated to be conclusive, the parties have the normal limitation periods of either six or 12 years in which to bring an action, depending on whether the contract is under hand or under seal respectively.

The conclusion is that, provided that you have left nothing or very little to be to your satisfaction or approval, and provided that the quantity surveyor has done his adjustment of the contract sum correctly, you have nothing to fear from issuing the final certificate.

11.7 Variations

Unless the contract is extremely simple, the valuation of variations is best left in the hands of the quantity surveyor. This attitude is adopted in IFC84, and clause 3.7 requires that the quantity surveyor shall make all valuations of variations. There is just one exception to this rule. The

Payment

Fig 11.1
Contract time chart

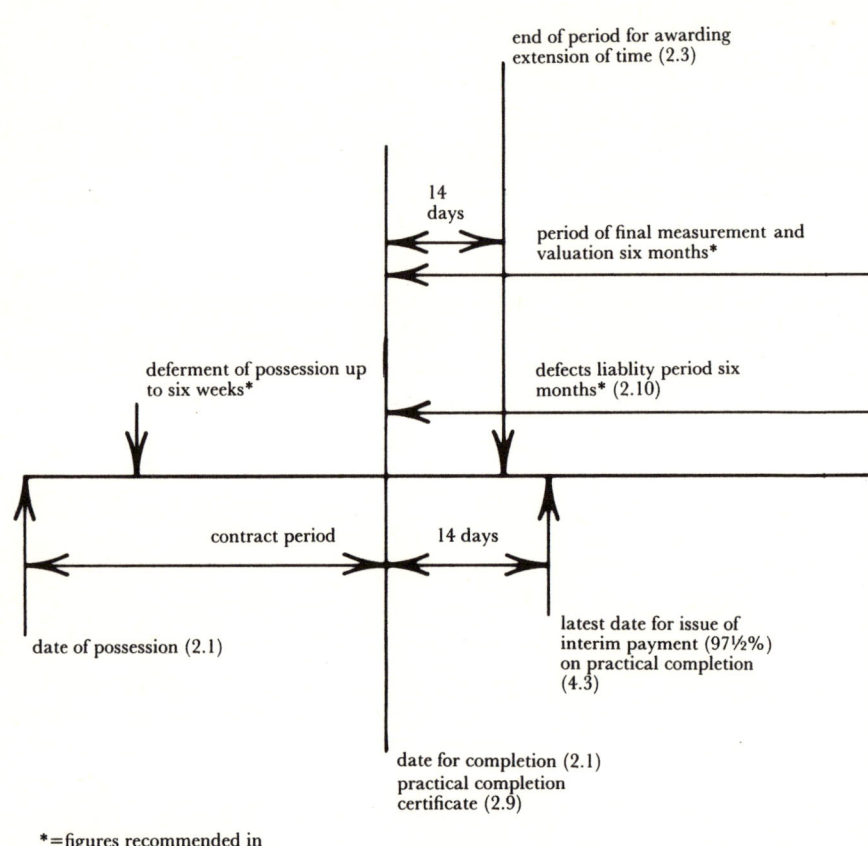

end of period for awarding extension of time (2.3)

14 days

period of final measurement and valuation six months*

deferment of possession up to six weeks*

defects liablity period six months* (2.10)

contract period

14 days

date of possession (2.1)

latest date for issue of interim payment (97½%) on practical completion (4.3)

date for completion (2.1)
practical completion certificate (2.9)

*=figures recommended in the contract

162

11.7 Variations

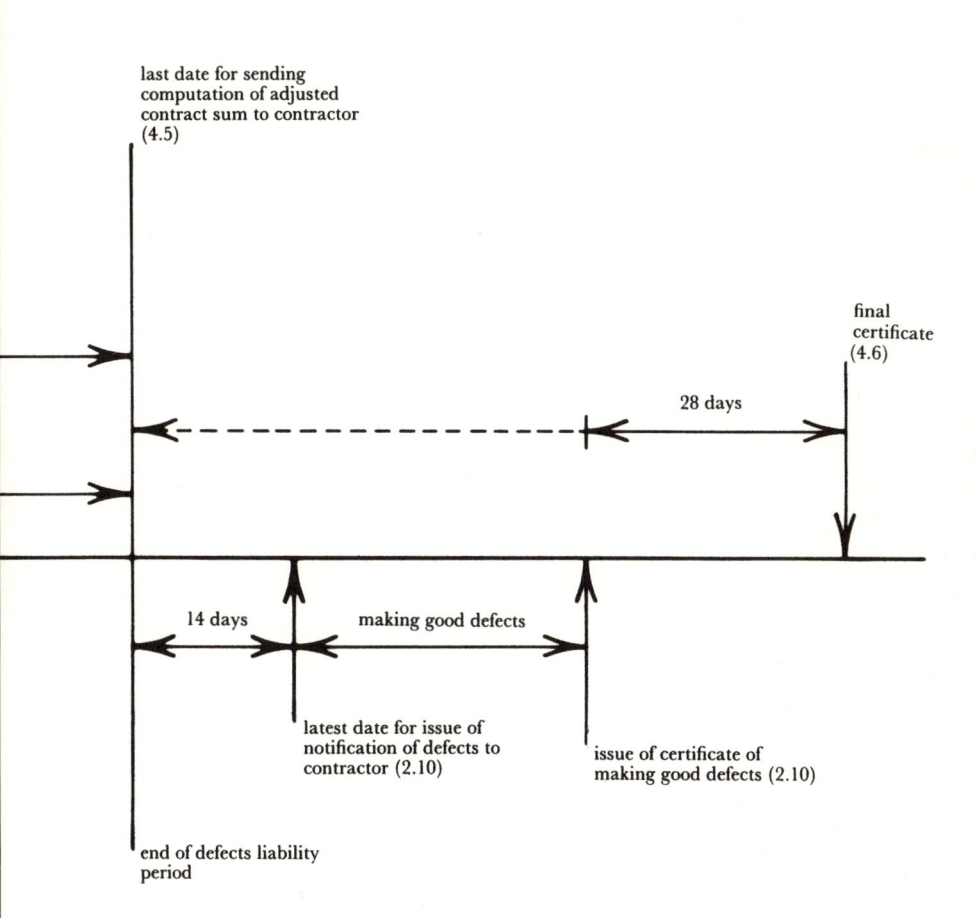

last date for sending computation of adjusted contract sum to contractor (4.5)

final certificate (4.6)

28 days

14 days

making good defects

latest date for issue of notification of defects to contractor (2.10)

issue of certificate of making good defects (2.10)

end of defects liability period

163

contractor and the employer may agree on the amount to be added to, or deducted from, the contract sum before the contractor carries out the work. This simply states the general position, because two parties to a contract can vary its terms in any way they both agree. The inclusion of the term merely makes it clear that if the contractor is asked to quote for carrying out some work which would normally be the subject of a valuation by the quantity surveyor, the employer may simply accept the quotation. The amount of the quotation is then added to, or deducted from, the contract sum as appropriate. Note that the contract expressly reserves the right of agreement to the employer, not to you.

The quantity surveyor's principal tool in carrying out valuations is what this contract refers to as the 'priced document' (clause 3.7.1). The wide range of possible documents has been discussed in section 3.1.1. The priced document may be any one of the following: the priced specification, or the priced schedule of work, or the priced bills of quantities, or the contract sum analysis, or the schedule of rates.

Omissions are to be valued in accordance with the relevant prices in the priced document. That is crystal clear and should cause no problems (clause 3.7.2).

Additional work may be valued in one of three ways:

— Work of similar character to that set out in the priced document: the valuation must be consistent with the values in the priced document, with due allowances for any change in the conditions or quantity. In this clause, it is probable that 'similar' can be given its ordinary meaning of 'almost identical', because the valuation is to be 'consistent'—ie, following the same principles—with the values in the priced document. Thus, if there is no change in the conditions under which the work is to be carried out and the quantity is unchanged, the prices in the document are to form the starting point for the valuation. Insofar as the 'similar' character is different, the values, to be consistent, must be different also. Where the conditions and quantity are also changed, the valuation will depart even further from the values in the priced document. There will clearly come a point at which the relationship with the priced document will be very tenuous indeed and, in effect, a fair valuation will result.

— If there is no work of similar character in the priced document, or to the extent that the execution of additional or substituted work or the omission of work is not concerned, or to the extent that work or liabilities directly associated with the instruction cannot be valued in the same way as work of a similar character, then a fair valuation must be made.

— Where the priced document is the contract sum analysis or the schedule of rates, if the prices set out are not relevant, a fair valuation must be made (clauses 3.7.4 and 3.7.9).

The procedure is set out in a flowchart (Fig 11.2). In practice, the

11.7 Variations

quantity surveyor will look at your instruction requiring a variation and see whether or not it is of similar character to work included in the priced document. If it is, he will use those prices as a basis for his valuation. If not, he will make a fair valuation. The wording of the clauses probably gives the quantity surveyor considerable freedom.

If it is decided that the proper basis of any fair valuation should be daywork, the valuation must comprise the prime cost of the work together with percentage additions on the prime costs at the rates set out by the contractor in the priced documents.

Definitions of prime costs are many and varied. The contract sets out those which are acceptable:

Generally: prime cost is to be calculated in accordance with the Definition of Prime Cost of Daywork carried out under a Building Contract, issued by the Royal Institution of Chartered Surveyors and the Building Employers' Confederation (formerly the National Federation of Building Trades Employers).

— If the work is within the province of any specialist trade and there is a published agreement between the RICS and the appropriate employers' body: prime cost is to be calculated in accordance with the definition in such an agreement. A footnote to clause 3.7.5 states that this sub-paragraph refers to three definitions, namely those agreed between the RICS and the Electrical Contractors' Association; between the RICS and the Electrical Contractors' Association of Scotland; and between the RICS and the Heating & Ventilating Contractors' Association. Since the footnotes are not part of the contract, there is nothing to prevent other definitions of prime cost from being used, provided that they fall within the meaning of the sub-paragraph.

The definitions to be used are those which were current at the date of tender.

There is no express requirement for the contractor to submit vouchers for verification, and it is for you and the quantity surveyor to decide what is required and to let the contractor know. Contractors generally prefer the valuation to be done by means of daywork, so there should be no difficulty. It is for the quantity surveyor, not you, to decide, in the light of the contract provisions, which method of valuation is most appropriate.

Additions or reductions to appropriate preliminary items must be included in the valuations (clause 3.7.6).

If the conditions under which any other work is carried out are substantially changed by reason of the contractor's carrying out of work in accordance with your instructions, the other work must be treated as if it too had been varied (clause 3.7.8). For example, your instruction to divide a large warehouse with brick walls may make it more difficult to lay floor screeds in some areas. In such a case, the quantity surveyor

Payment

Fig 11.2
Valuation of variations (clause 3.7)

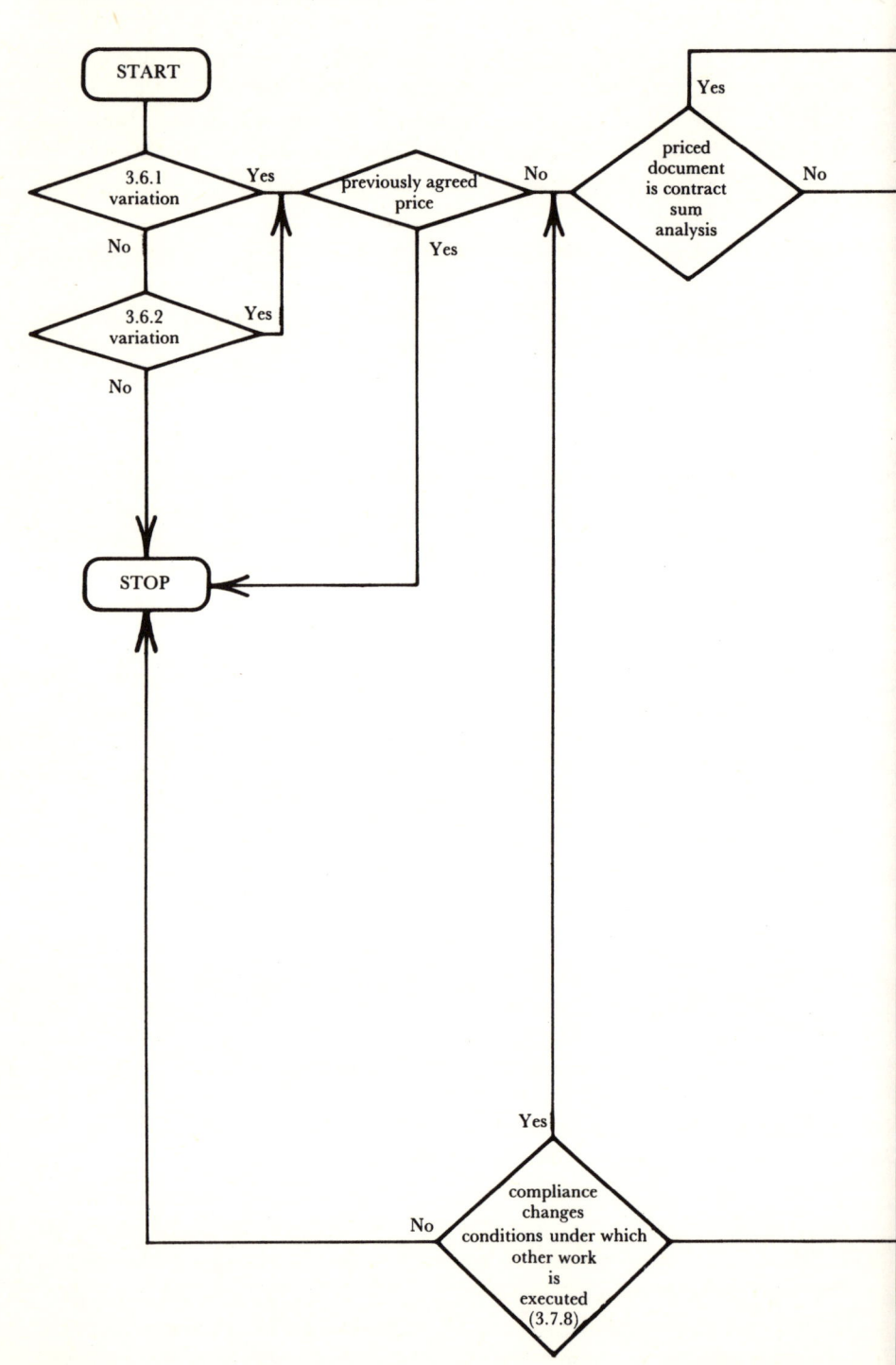

11.7 Variations

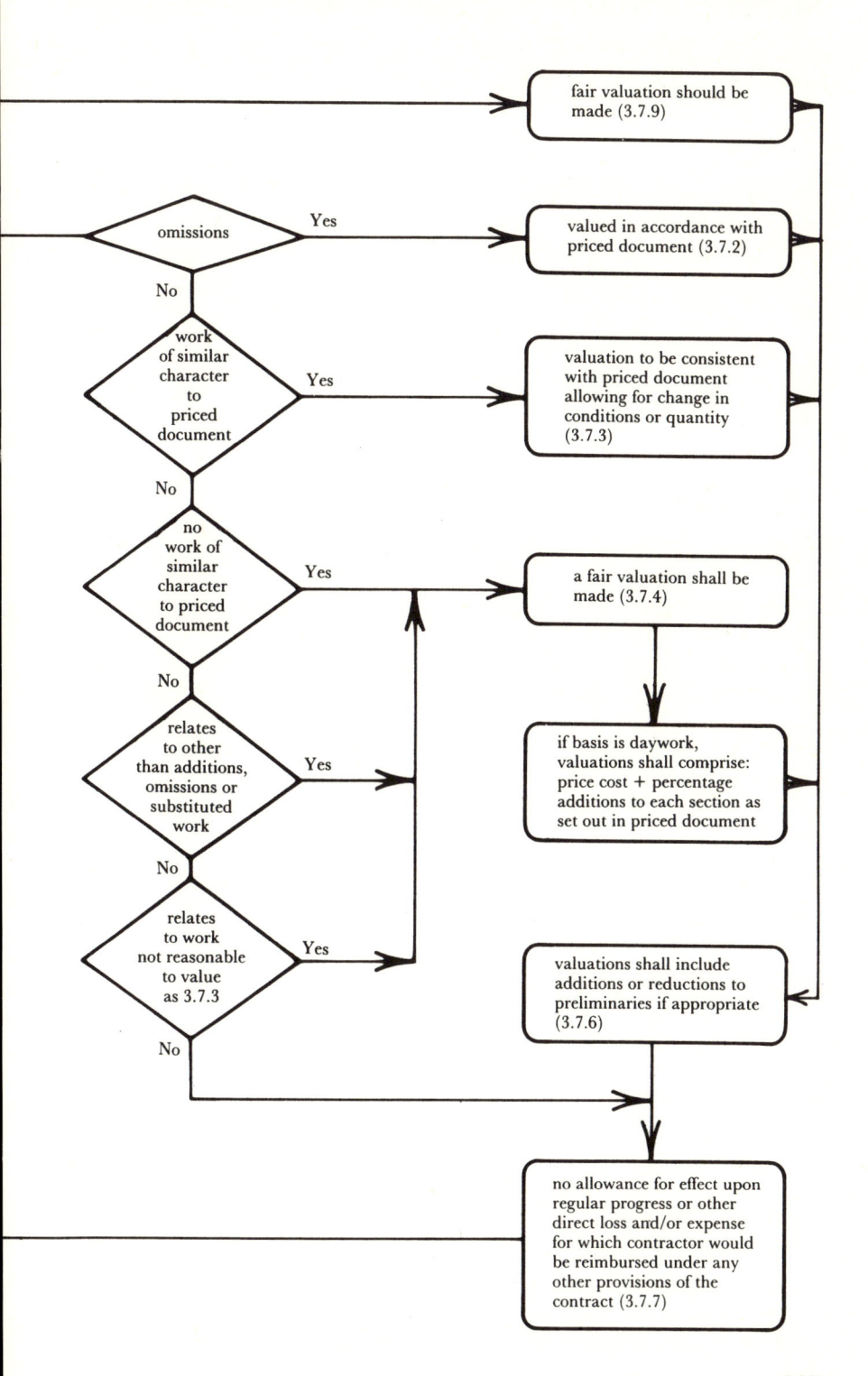

```
                                          ┌─────────────────────┐
                                      ──▶ │ fair valuation should be
                                          │ made (3.7.9)         │
                                          └─────────────────────┘

         ◇ omissions ──Yes──▶ ┌─────────────────────┐
              │               │ valued in accordance with
              No              │ priced document (3.7.2)
              ▼               └─────────────────────┘
         ◇ work of similar
           character to      ──Yes──▶ ┌─────────────────────┐
           priced document            │ valuation to be consistent
              │                       │ with priced document
              No                      │ allowing for change in
              ▼                       │ conditions or quantity (3.7.3)
                                      └─────────────────────┘
         ◇ no work of similar
           character to      ──Yes──▶ ┌─────────────────────┐
           priced document            │ a fair valuation shall be
              │                       │ made (3.7.4)
              No                      └─────────────────────┘
              ▼
         ◇ relates to other
           than additions,  ──Yes──▶ ┌─────────────────────┐
           omissions or              │ if basis is daywork,
           substituted work          │ valuations shall comprise:
              │                      │ price cost + percentage
              No                     │ additions to each section as
              ▼                      │ set out in priced document
         ◇ relates to work           └─────────────────────┘
           not reasonable  ──Yes──▶ ┌─────────────────────┐
           to value as 3.7.3        │ valuations shall include
              │                     │ additions or reductions to
              No                    │ preliminaries if appropriate
              ▼                     │ (3.7.6)
                                    └─────────────────────┘
                                    ┌─────────────────────┐
                                    │ no allowance for effect upon
                                    │ regular progress or other
                                    │ direct loss and/or expense
                                    │ for which contractor would
                                    │ be reimbursed under any
                                    │ other provisions of the
                                    │ contract (3.7.7)
                                    └─────────────────────┘
```

must value not only the cost of building the brick walls but also the cost of laying the floor screeds under different conditions. The change must be substantial—ie, it must not be trivial. It is for the quantity surveyor to decide where to draw the line.

The quantity surveyor is not to make any allowance in his valuation for the effect of your instruction on the regular progress of the works or for any other direct loss and/or expense for which the contractor would be reimbursed by any other provision of the contract (clause 3.7.7). This is clearly intended to avoid confusion between this clause and clause 4.11 and to prevent the contractor from claiming, and being paid, twice for the same matter. It is still open, of course, for the contractor to satsify the quantity surveyor that the effect on progress and loss and/or expense are not covered by other contract provisions. If he can, he is entitled to be paid under this clause.

11.8 Fluctuations

There are two clauses which deal with fluctuations, clauses 4.9 and 4.10. The first deals with fluctuations allowable to the contractor's work; the second with fluctuations in respect of named persons. The fluctuations referred to in clause 4.9 are contained in supplemental conditions C and D. They are published as a separate booklet entitled JCT Fluctuations Clauses for use with the JCT Intermediate Form of Building Contract IFC84.

Supplemental condition C will always apply unless supplemental condition D is stated in the appendix to apply. It is the contract sum, less amounts included for work by named persons, which is to be adjusted.

Supplemental condition C, contribution, levy, and tax fluctuations, provides for the bare minimum of fluctuations to cover changes in statutory payments such as National Insurance contributions.

Supplemental condition D, use of price adjustment formulae, is applicable only if bills of quantities are included in the contract documents. It allows for fluctuations in accordance with formula rules published by the Joint Contracts Tribunal. In effect, this provides full fluctuations.

The amounts included in the contract sum in respect of named persons are to be adjusted by the net amounts which the named person is due to receive or allow in accordance with the applicable subcontract (NAM/SC) fluctuation provisions 33 or 34. There is a proviso. It applies if the period for completion of the subcontract works has been extended by reason of an act, omission, or default of the contractor (NAM/SC clause 12.2.1). In that case, any sum which would have been excluded, were it not for the extension of time, is to be excluded from the net

11.8 Fluctuations

amount mentioned above. This is because NAM/SC clause 33.4.7 or 34.7.1 (as applicable) 'freezes' the application of the fluctuations clause if the subcontractor fails to complete on time. Therefore, if it is the contractor's fault that the subcontractor failed to complete, he is still liable to pay the fluctuations to the subcontractor, but the employer need not pay those amounts to the contractor. The subcontract provisions are complex. For example, if the NAM/SC provisions for extension of time are amended in any way, or if the contractor does not respond to claims for extension of time within the prescribed period, the right to freeze fluctuations is lost (NAM/SC clauses 33.4.8, and 34.7.2 and 3). It is thought that in such circumstances, the proviso to clause 4.10 would be robbed of all effect, whether the amendment was carried out by the employer or by the contractor.

11.9 Summary

Contract sum

— the contract sum is the amount for which the contractor agrees to carry out the whole of the work
— IFC84 is a 'lump-sum' contract
— if the contractor abandons the work, he is entitled to nothing
— the contract sum can be adjusted only in accordance with the provisions of the contract
— once the contractor's price has been accepted, it is fixed.

Payment before practical completion

— the parties may make their own arrangements
— if there is no agreement to the contrary, clause 4.2 applies
— the usual interval is one month between certificates
— the employer has 14 days in which to pay
— a valuation may be carried out by the quantity surveyor not more than seven days before the certificate
— you should decide what is meant by 'value' and inform the contractor at the time of tender
— the employer is not allowed a retention on all amounts.

Payment at practical completion

— the certificate must be issued within 14 days of the date of practical completion
— half the retention must be released.

Payment

Retention

— unless the employer is a local authority, he acts as trustee for the retention fund
— the contractor has the right to require the retention fund to be kept in a separate bank account
— the employer, if not a local authority, has the right to take out money to pay amounts which the contract provides for him to deduct
— the position of a local authority is unclear.

Final payment

— the contractor must send all documents for computing the final adjusted contract sum
— they must be received before, or a reasonable time after, practical completion
— a copy of the quantity surveyor's final computations must be sent to the contractor before the end of the period of final measurement and valuation
— the final certificate must be issued 28 days from either the date the computations are sent, or the date of the certificate of making good defects, whichever is the later.

Effect of certificates

— no certificate is conclusive evidence that work or materials are in accordance with the contract
— the final certificate is conclusive that, if you are to be satisfied, you are satisfied; and that terms of the contract requiring additions, deductions, or adjustments to the contract sum have been correctly operated
— the final certificate is not conclusive if items have been accidentally included or omitted or there are arithmetrical errors, or if either party commences proceedings before 21 days after the date of issue.

Variations

— the amount may be agreed between the employer and the contractor before compliance
— otherwise, the quantity surveyor must value in accordance with clause 3.7.

11.9 Summary

— valuation may be based on the priced document or be a fair valuation
— the basis for a fair valuation may be daywork
— acceptable definitions of 'prime cost' are indicated in clause 3.7.5
— adjustment to preliminary items must be included
— work changed by a variation to other work must be treated as a variation
— there is no allowance for loss and/or expense unless the contractor cannot otherwise recover.

Fluctuations

— supplemental condition C applies unless D is stated in the appendix to apply
— C covers bare minimum statutory changes
— D covers full fluctuations
— named person fluctuations are to be paid net
— if subcontract works are extended through the contractor's fault, no fluctuations are paid for that period.

12 Determination

12.1 Determination by the employer

12.1.1 General

Determination is one of those things which are best avoided. If it is impossible to avoid, it must be done properly, or the consequences will be unpleasant for the employer. The whole procedure is surrounded by difficulties and pitfalls for the unwary. Among them are the following.
If determination is properly carried out, the employer will be faced with a project to finish with the aid of another contractor. In theory, he can recover all his costs from the first contractor, but he cannot recover the time lost. If determination is not properly carried out, the contractor may be able to bring an action for damages for unlawful repudiation of the contract. And many of the grounds for determinatioin may give rise to dispute.
The employer will look to you for advice on whether or not to determine the contractor's employment. Indeed, it will probably be you who brings the matter to the attention of the employer. This will usually be because the situation on site has deteriorated to such a stage that you are pessimistic about the chances of ever achieving completion. Ideally, you should have set the determination in motion before you get to that stage, but in practice it is difficult to decide just when there is no hope of recovering the situation.
The procedure for determination is set out in a flowchart (Fig 12.1). The grounds for determination are set out in the contract in clauses 7.1, 7.2, 7.3, 7.8.1, and 6.3C.2. The consequences of determination are set out in clauses 7.4, 7.9, and 6.3C.2.

12.1 Determination by the employer

12.1.2 Grounds (clause 7.1)

There are four separate grounds for determination in clause 7.1. They are that the contractor:
— wholly suspends the carrying out of the work, before completion, without reasonable cause; or
— fails to proceed regularly and diligently with the works; or
— refuses or persistently neglects to comply with a written notice from the architect requiring him to remove defective work or improper materials or goods, and thereby the works are materially affected; or
— fails to comply with clauses 3.2 (subcontracting), 3.3 (named persons), or 5.7 (fair wages).

If the employer, with your advice, decides to determine the contractor's employment, you must ensure that the procedure is followed precisely. The contractor must be served with a notice of default (letter, Fig 12.2), which must clearly specify the default. The contract is silent on who is to send the letter, so it is prudent to draft a letter for the employer to sign. The letter must be sent by registered post or recorded delivery. It may be thought that delivering by hand and obtaining a receipt would be just as good or even more certain, but it is wise to follow the contract provisions. If the contractor continues the default for 14 days after receipt of the notice, the employer may determine the contractor's employment by a further notice by registered post or recorded delivery. You should draft the letter for the employer's signature (Fig 12.3).

A point sometimes arises concerning the date on which the first notice was received and thus from which the 14 days begin to run. You can make an assumption about the date on which the notice would arrive in the ordinary course of the post, but if you are wrong and the contractor can prove that the notice of determination was premature, the employer may be taken to have repudiated the contract unlawfully. The wise course is to arrange for the Post Office to confirm the delivery date to you.

There are two important provisos. First, if the contractor ceases his default within the 14 days, the employer can take no immediate action. But if the contractor repeats the *same default* at any time thereafter, the employer may determine forthwith without the necessity for a further 14 days' notice. This is a very powerful remedy in the hands of the employer. Second, the notice of determination must not be given unreasonably or vexatiously. There must be no malice or intention to annoy. This is particularly applicable to the case where the contractor has stopped a default for some weeks or months but commits the same default again. You must be specially careful that a change of unreasonableness cannot successfully be levelled at the employer. Despite the provisions of the contract, it would be prudent to send a

173

Determination

Fig 12.1
Determination by employer (clauses 7.1, 7.8, 6.3C.2)

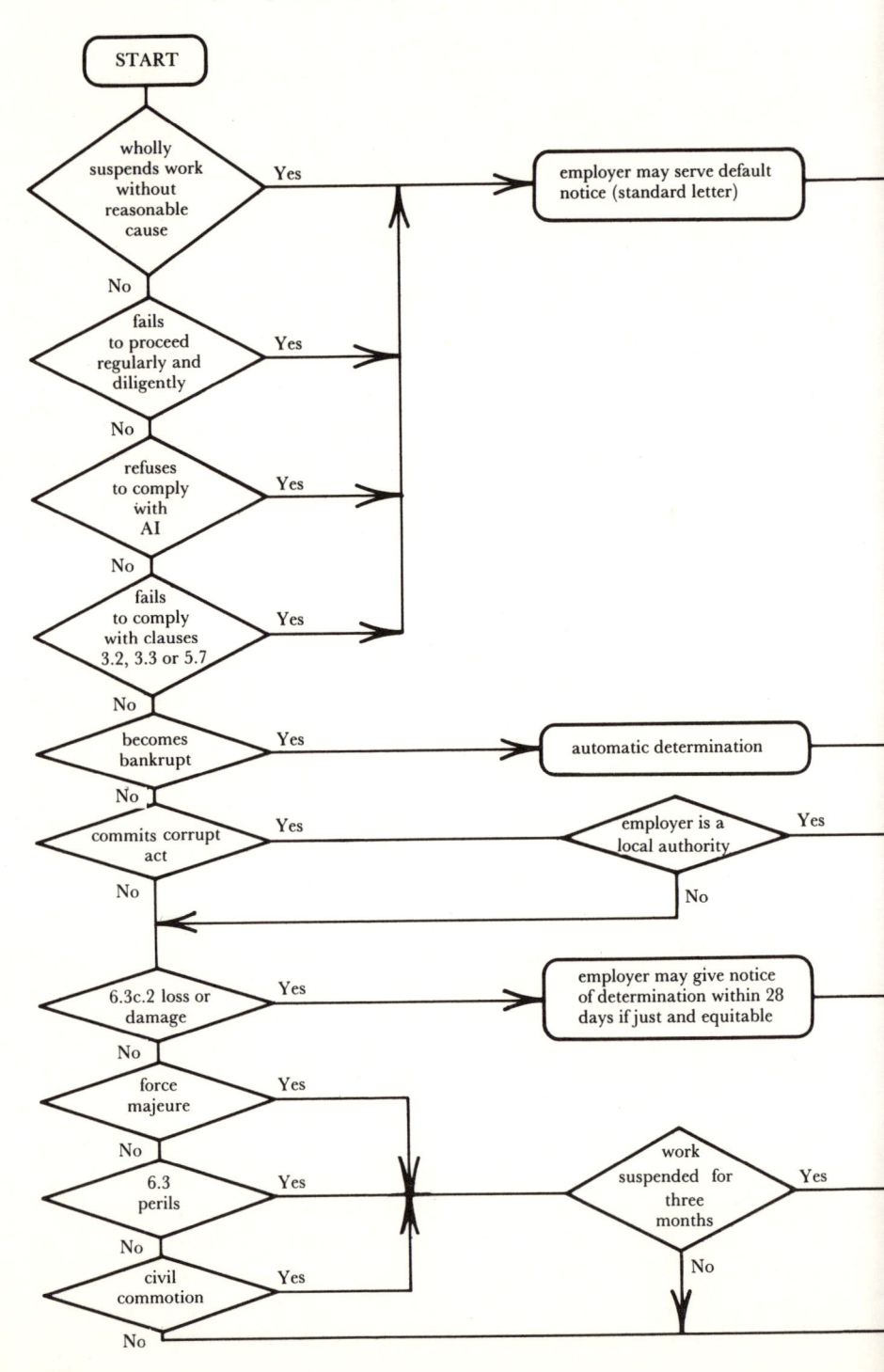

```
                    START
                      │
            ┌─────────▼─────────┐
            │ wholly suspends   │   Yes    ┌──────────────────────────┐
            │ work without      ├────────► │ employer may serve       │
            │ reasonable cause  │          │ default notice           │
            └─────────┬─────────┘          │ (standard letter)        │
                     No                    └──────────────────────────┘
            ┌─────────▼─────────┐
            │ fails to proceed  │   Yes
            │ regularly and     ├────────►
            │ diligently        │
            └─────────┬─────────┘
                     No
            ┌─────────▼─────────┐
            │ refuses to comply │   Yes
            │ with AI           ├────────►
            └─────────┬─────────┘
                     No
            ┌─────────▼─────────┐
            │ fails to comply   │   Yes
            │ with clauses      ├────────►
            │ 3.2, 3.3 or 5.7   │
            └─────────┬─────────┘
                     No
            ┌─────────▼─────────┐   Yes    ┌──────────────────────────┐
            │ becomes bankrupt  ├────────► │ automatic determination  │
            └─────────┬─────────┘          └──────────────────────────┘
                     No
            ┌─────────▼─────────┐   Yes    ┌──────────────────┐  Yes
            │ commits corrupt   ├────────► │ employer is a    ├────►
            │ act               │          │ local authority  │
            └─────────┬─────────┘          └────────┬─────────┘
                     No                            No
            ┌─────────▼─────────┐   Yes    ┌──────────────────────────┐
            │ 6.3c.2 loss or    ├────────► │ employer may give notice │
            │ damage            │          │ of determination within  │
            └─────────┬─────────┘          │ 28 days if just and      │
                     No                    │ equitable                │
            ┌─────────▼─────────┐   Yes    └──────────────────────────┘
            │ force majeure     ├────────►
            └─────────┬─────────┘
                     No                            ┌──────────────────┐  Yes
            ┌─────────▼─────────┐   Yes            │ work suspended   ├────►
            │ 6.3 perils        ├────────►         │ for three months │
            └─────────┬─────────┘                  └────────┬─────────┘
                     No                                    No
            ┌─────────▼─────────┐   Yes                    ▼
            │ civil commotion   ├────────►
            └─────────┬─────────┘
                     No
```

174

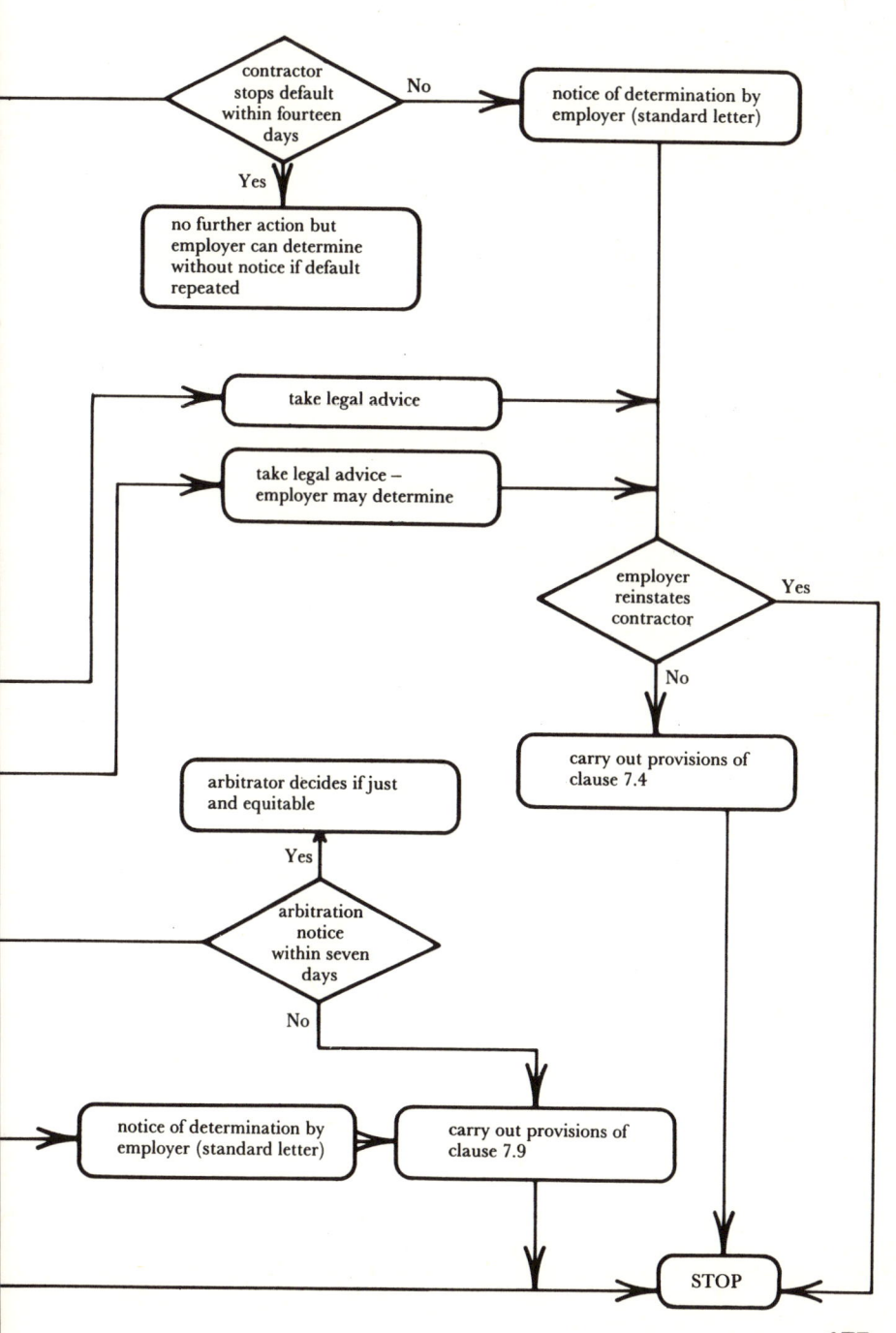

12.1 Determination by the employer

- contractor stops default within fourteen days?
  - **Yes** → no further action but employer can determine without notice if default repeated
  - **No** → notice of determination by employer (standard letter)

- take legal advice
- take legal advice – employer may determine

- employer reinstates contractor?
  - **Yes** → STOP
  - **No** → carry out provisions of clause 7.4

- arbitration notice within seven days?
  - **Yes** → arbitrator decides if just and equitable
  - **No** → notice of determination by employer (standard letter) → carry out provisions of clause 7.9 → STOP

Determination

Fig 12.2
Employer to contractor, giving notice of default

REGISTERED POST OR RECORDED DELIVERY

Dear Sir

I hereby give you notice under clause 7.1 of the conditions of contract that you are in default in the following respect: [*insert details of the default, with dates if appropriate*].

If you continue the default for 14 days after receipt of this notice, or if you at any time repeat such default (whether previously repeated or not), the employer may thereupon determine your employment under this contract without further notice.

Yours faithfully

Copy: Architect
      Quantity surveyor

12.1 Determination by the employer

Fig 12.3
Employer to contractor, determining employment

```
REGISTERED POST OR RECORDED DELIVERY


Dear Sir

I refer to the notice dated  [insert date of
original notice].

In accordance with clause 7.1 of the conditions of
contract, take this as notice that I hereby
determine your employment under this contract
without prejudice to any other rights or
remedies which I may possess.

The rights and duties of the parties are governed
by clause 7.4.  The architect will write to you
within the next seven days with instructions
regarding the temporary buildings, plant, tools,
equipment, goods and materials on site.  Subject
only to your compliance with the architect's
instructions you must give up possession of the
site forthwith.


Yours faithfully


Copy: Architect
      Quantity surveyor
```

Determination

Fig 12.4
Architect to contractor, giving warning of repeated default

Dear Sir

This is not a notice of default under clause 7.1 of the conditions of contract.

A notice of default dated [*insert date*] was sent to you in respect of the following: [*insert details of the default*].

It has come to my attention that the above default is being repeated. Under clause 7.1, the employer has the right to determine your employment under this contract without further warning. The employer intends to exercise his right unless the default ceases immediately. I will visit the site on [*insert date*], and if you are still in default, I will advise the employer accordingly.

Yours faithfully

Copy: Employer
      Quantity surveyor

12.1 Determination by the employer

warning letter (Fig 12.4), being careful to state that it is not a notice of default, or you will start the 14 day period again. This lets the contractor know that the employer intends to exercise his right to determine. Usually, that will be sufficient to stop the default immediately. If it does not, the employer would certainly not be acting unreasonably if he then determined the contractor's employment.

Before you advise the employer to give notice, consider the four grounds carefully.

'Wholly suspends the carrying out of the work', etc (clause 7.1(a)):

The contractor must have completely ceased work, which probably means that he will have left the site. If a contractor, half way through a million-pound contract, has only one or two men on site doing token work, he would probably be regarded as having wholly suspended the work, but it is by no means certain. This ground appears to be intended to cover the situation where the contractor has, in effect, abandoned the work. Note that the suspension must be without reasonable cause. Before advising the employer to send the initial notice of default, you would have to ask the contractor why he had stopped. Your course of action would depend on his reply. He might, for example, have suspended work for some reason which ranked for an extension of time.

'Fails to proceed regularly and diligently with the works' (clause 7.1(b)):

This ground implies more than simply failing to keep to his programme. The contractor's programme is not a contract document, although it may be a good indication of the contractor's intentions. 'Regularly and diligently' means that the contractor must work constantly, systematically, and industriously. In deciding whether the contractor is working as required, regard must be had to such things as these: the number of men on site, compared to the number of men required; the amount of plant and equipment in use; the work to be done; the time available for completion of the work; the actual progress being made; and factors outside the contractor's control (which may not all be clause 2.3 'events') which hinder progress. It is clear that it is no easy matter to prove lack of regular and diligent progress. It has been suggested that a contractor will be able to argue his way out of any determination on this ground if he is making any progress at all.

'Refuses or persistently neglects to comply', etc (clause 7.1(c)):

Refusal to comply with the architect's instructions to remove defective work can be dealt with under clause 3.5.1, the employer engaging others to do the work. This ground is probably intended to cover the situation when the contractor ignores such instructions to such an extent that the work is in danger of grinding to a halt. It would also refer to the case where so much work is defective that further work cannot be done without 'building in' the defective work and necessitating any future satisfactory work being taken down to make good the defective parts. The work on site would have to be in a very sorry state before this ground would apply.

Fails to comply with clauses 3.2, 3.3, or 5.7 (clause 7.1(d)):

Clause 3.2 refers to subcontracting without consent. The object is to prevent the contractor from arranging vicarious performance of part of the contract by another. Because subcontracting is traditional in the building industry, the contract makes provision for it provided that the architect consents. If the contractor does sublet without consent, it is for you to decide whether the subcontractor is suitable. If he is, there would seem to be no point, and little chance of success, in trying to determine the contract. Even if the subcontractor is clearly unsuitable, a less draconian method of dealing with the situation, probably by a letter (Fig 12.5), would probably be appropriate. Determination is best reserved for the occasions when the contractor sublets the whole or large portions of the work without your approval.

Clause 3.3 refers to named persons. The provisions of this clause are complex but, briefly, this ground is aimed at giving the employer a remedy if the contractor fails to subcontract with a named person under any of the procedures or is otherwise in breach of his obligations under that clause.

Clause 5.7 refers to fair wages. It is applicable only if the employer is a local authority. Since the Fair Wages Resolution of the House of Commons was rescinded in 1983, it is extremely doubtful whether an attempted determination on this ground would be successful.

12.1.3 Grounds (clause 7.2)

Under this clause, determination is automatic for any of the following reasons:
— the contractor becomes bankrupt
— he makes a composition or arrangement with his creditors
— he has a winding up order made (except for the purposes of amalgamation or reconstruction)
— a resolution for voluntary winding up is passed
— a provisional liquidator, receiver, or manager of the contractor's business is duly appointed

12.1 Determination by the employer

Fig 12.5
Architect to contractor if contractor sublets without consent

Dear Sir

It has been brought to my attention that you have purported to sublet [*insert the portion of the works sublet*] to [*insert name of purported subcontractor*].

You have taken this course of action after I have refused consent/without asking my consent [*delete as appropriate*] to subletting. Since I have no intention of giving my consent, you are not permitted to use the above-mentioned purported subcontractor on the works.

I send this letter because I trust the incident is an oversight on your part and I am reluctant to advise the employer to use his rights under clause 7.1 if you will confirm to me, by return, that you will comply with this letter.

Yours faithfully

Copy: Employer
      Quantity surveyor
      Clerk of works

Determination

— possession is taken, by or on behalf of the holders of any debentures secured by a floating charge, of any property comprised in, or subject to, the floating charge

In general terms—if the contractor becomes insolvent.

Neither you nor the employer is required to do anything further if you wish the determination to stand. However, even when a receiver or manager is appointed, it may be in the best interests of the employer to continue the contract. You will be in a position to decide this only after a thorough discussion with all parties concerned. It is also prudent to obtain, through the employer, legal advice regarding the implications in a particular case. The contractor's employment may be reinstated if agreement can be reached between the employer and the contractor, his trustee in bankruptcy, liquidator, provisional liquidator, receiver, or manager (whichever is appropriate).

12.1.4 Grounds (clause 7.3)

This clause applies only if the employer is a local authority. The employer may determine the contractor's employment if the contractor has given or received bribes in connection with this or any other contract with the employer, or if the contractor commits any other offence in relation to the contract or any other contract with the employer under the Prevention of Corruption Acts 1889 to 1916 or under subsection (2) or s117 of the Local Government Act 1972. A most onerous part of the provisions of this clause, so far as the contractor is concerned, is the fact that his employment may be determined because of the corrupt actions of one of his employees or of some person acting on his behalf. It matters not that the contractor may have no knowledge of the affair.

In any case, corruption is a criminal offence for which there are strict penalties, and the employer is entitled at common law to rescind the contract and/or recover any secret commissions.

Legal advice is indicated, followed by a simple notice of determination, if that is the decision.

12.1.5 Grounds (clause 7.8.1)

The employer (or the contractor) may determine the contractor's employment if the carrying out of the whole or substantially the whole of the uncompleted works is suspended for three months because of *force majeure*, or loss or damage to the works caused by clause 6.3 perils or civil comotion.

For either party to operate this clause, the works must be totally lacking in any significant progress for the entire three-month period as a result of the same cause. One week of frenzied activity on the part of the

12.1 Determination by the employer

contractor, after one month of the period has elapsed, may be sufficient to prejudice any attempt at determination, even if the site relapses into inactivity for two months thereafter. It is probable, however, that the period must be viewed as a whole.

No period of notice is required. At the end of the three months suspension of work, either party may forthwith determine the contractor's employment by written notice sent by registered post or recorded delivery (letter Fig. 12.6). The clause contains a proviso that the notice must not be given unreasonably or vexatiously.

All the causes of suspension are events beyond the control of the parties. *Force majeure* (discussed earlier) could embrace a war, a major strike, and any violent disturbance down to a civil commotion (the third of the causes). A civil commotion is more serious than a riot but not as serious as a civil war. Clause 6.3 perils are the insurance risks noted earlier.

12.1.6 Grounds (clause 6.3C.2)

This clause provides for the employer (or the contractor) to determine the contractor's employment within 28 days of the occurrence of loss or damage to the works or to any unfixed materials or goods caused by any clause 6.3 perils. The clause refers to existing structures to which work is being done by way of alteration or extension or both. The contractor must give notice in writing to you and to the employer as soon as he discovers the damage. His notice must state the extent, nature, and location of the damage. Although the 28 days begin to run from the occurrence and not from the notification, in practice, if the damage is likely to be such as to form the basis for determination, it will be discovered and notified immediately it occurs.

A very important proviso says, 'if it is just and equitable to do so'. This proviso goes to the heart of the matter and points the difference between this ground for determination and the ground in clause 7.8.1(b), which requires a three-month period of suspension. What is just and equitable is dependent on the particular circumstances. Despite the words at the beginning of the clause, 'If any loss or damage', the sort of situation in which determination would clearly be just and equitable involves such catastrophic damage that is it uncertain not only when work could recommence but whether work could recommence at all.

Take the case of a large factory, worth several million pounds. It may be that a small alteration and extension contract is let, worth £150,000. If, during the course of the work, the whole building is totally destroyed by fire, it will be just and equitable to determine the contractor's employment. (In that case, the contract may also be considered to be

183

Fig 12.6
Employer to contractor, determining employment under clause 7.8.1

REGISTERED POST OR RECORDED DELIVERY

Dear Sir

The whole or substantially the whole of the uncompleted works has been suspended since [*insert date*], a period of three months, by reason of: [*insert reason for suspension*].

In accordance with clause 7.8.1 of the conditions of contract, take this as notice that I forthwith determine your employment under this contract without prejudice to any other rights or remedies which I may possess.

The rights and duties of the parties are governed by clause 7.9. The architect will draw up a statement of account as soon as reasonably practicable.

Yours faithfully

Copy: Architect
      Quantity surveyor

12.1 Determination by the employer

frustrated.) Even much less than total destruction in such circumstances would give grounds for determination under this clause.

The right of either party to seek arbitration is limited in two ways. Written request to concur in the appointment of an arbitrator under article 5 must be given within seven days of receipt of a notice of determination, and the arbitrator is to decide whether determination will be just and equitable.

12.1.7 Consequences (clause 7.4)

This clause lays down the procedure to be followed after determination under clauses 7.1, 7.2, and 7.3. The procedure is stated to be 'without prejudice to any arbitration or proceedings in which the validity of the determination is in issue'. In other words, the procedure cannot affect the arbitration and the determination itself cannot become a *fait accompli*. In the event that the contractor seeks arbitration, the employer may deem it prudent to await the outcome in any case, because, if it is not possible to reinstate the original contractor, he will be entitled to substantial damages (including loss of profit) if he wins.

Assuming that the contractor does not seek arbitration or that the employer decides, nevertheless, to proceed, the consequences of the determination are as follows:

— the contractor must give up possession of the site of the works. If he does not do so within a reasonable time after receiving the notice of determination, he will become, in law, a trespasser, and you should send him a further notice to that effect (letter, Fig 12.7). The contractor's liability for insurance ceases, and you must ensure that the employer takes out appropriate insurance cover without delay. This is best done at the time the notice of determination is sent (letter, Fig 12.8). This clause makes no provision for the contractor to ensure that the works are left in a safe condition, but the contractor (like anyone else) has a duty of care to those he can reasonably foresee could suffer injury. He must not, therefore, leave walls or beams in a precarious condition.

—unlike the equivalent clause in JCT80, there is nothing in this clause to prevent the contractor from removing all his temporary buildings, plant, goods, and materials, bought or hired by him, from site. Clause 7.4(c) allows the employer to use any temporary buildings, etc, for the benefit of a subsequent contractor and to purchase any other materials necessary for completing the works. As it stands, the clause appears to mean that the employer may use any temporary buildings, etc, which the original contractor has left. You must advise the employer whether use of the contractor's plant is desirable and may constitute a saving. If you decide not to use it, you have the power to instruct the original

Determination

Fig 12.7
Architect to contractor if contractor refuses to give up possession of site

REGISTERED POST OR RECORDED DELIVERY

Dear Sir

[*Either*]
The employer determined your employment under this contract under clause [*insert clause 7.1 or 7.3*] on [*insert date*].

[*Or*]
Your employment was automatically determined under this contract under clause 7.2 on [*insert date*].

[*Then*]
The consequences of determination are laid down in clause 7.4 Subsection (a) of that clause requires you to give up possession of the site. It is now [*insert number*] days since notice of determination was served on you and you have not given up possession. In law, you are a trespasser, and if you have not given up possession of the site by [*insert date*], the employer intends to take whatever action he deems necessary to secure your removal.

Yours faithfully

Copy: Employer

12.1 Determination by the employer

Fig 12.8
Architect to employer, regarding insurance if contractor's employment determined

Dear Sir

Your notice of determination is being sent to the contractor today. The contractor no longer has any liability to insure the works. You should consult your own broker without delay to obtain cover similar to that which the contractor was required to have under clauses 6.2 and 6.3A of the conditions of contract. Your insurance cover should be maintained at least until suitable arrangements have been made to complete the contract with another contractor.

Yours faithfully

Determination

contractor to remove it from the works. He has a reasonable time in which to comply, and then the employer may remove and sell it. All proceeds, less costs, must be held to the credit of the contractor.

— The employer may employ another contractor to complete the works. The completion of a partly finished building by another contractor is always an expensive procedure. In order to avoid a potential dispute, it is wise to have bills of quantities prepared for the completion work and to go out to tender in the normal way. This will prevent the original contractor from contending that the employer has not obtained a reasonable price.

Until the works are complete, the employer is not bound, and would not be wise, to make any further payment to the original contractor. When the contract is completed and the subsequent contractor paid in full, the employer must draw up a set of accounts which must show

— all the expenses and direct loss and/or damage caused to the employer by the determination. It will include the cost of completing the contract, including all professional fees consequent on the determination, and the cost of engaging another contractor.

— the amount paid to the original contractor before determination. If the total of the two amounts is greater or less than what would have been paid had the contract been completed in the normal way, the difference is a debt payable by the original contractor to the employer, or vice versa.

Invariably, the contractor owes a debt to the employer. You must be in no hurry to calculate, with the assistance of the quantity surveyor, the final amounts. It is essential that you make sure that, on paper at least, you have put the employer in the same position as he would have been had the contract not been determined but had continued in an orderly way to its conclusion. Obtaining payment from a contractor who may be insolvent is another matter (clause 7.4(d)).

12.1.8 Consequences (clauses 7.9 and 6.3C.2)

The consequences of determination covered by these clauses are covered in section 12.2.7.

12.2 Determination by the contractor

12.2.1 General

If the contractor is successful in determining his employment under the contract, the results for the employer will be catastrophic. So you must do all in your power to prevent this from happening. Among the consequences are these.

12.2 Determination by the contractor

— the employer will be left with the project to complete with another contractor, completion bills must be prepared, and a great deal of additional expense will be incurred in the form of increased cost of completion and additional professional fees. The employer will be looking around to blame, and possibly sue, someone. Make sure it is not you

— the completion date will be considerably exceeded

— under some of the grounds for determination, the contractor is entitled to receive loss of the profit he expected to make on the whole contract.

The procedure for determination by the contractor is set out in a flowchart (Fig 12.9). The grounds for determination are set out in clauses 7.5, 7.6, 7.8.1, and 6.3C.2. The consequences of determination are set out in clauses 7.7, 7.9, and 6.3C.2.

12.2.2 Grounds (clause 7.5)

There are three separate grounds for determination in clause 7.5. They are the following:

— the employer does not pay the contractor under clauses 4.2, interim payments; or 4.3, interim payments on practical completion; or 4.6, final certificate

— the employer interferes with, or obstructs, the issue of any certificate

— the carrying out of the whole or substantially the whole of the works is suspended for a continuous period of one month due to one of the following reasons:

— architect's instructions under clauses 1.4, inconsistencies; or 3.6, variations; or 3.15, postponement, unless caused by contractor's default

— the contractor not having received, at the proper time, instructions, etc, for which he specifically applied at a time reasonable in relation to the date on which they were required

— failure or delay in executing work not forming part of this contract by the employer or his men, or failure or delay in supplying materials which the employer has agreed to supply

— failure of the employer to give, at the appropriate time, ingress to, or egress from, the site or any part; this includes necessary passage over adjoining land in the possession and control of the employer after the contractor has given any notice required by the contract documents; or failure of the employer to give ingress or egress as agreed between the architect and the contractor.

If the contractor decides to determine on any of the above grounds, he must follow the procedure precisely, otherwise he may be simply

Determination

Fig 12.9
Determination by contractor (clauses 7.5, 7.8, 6.3C.2)

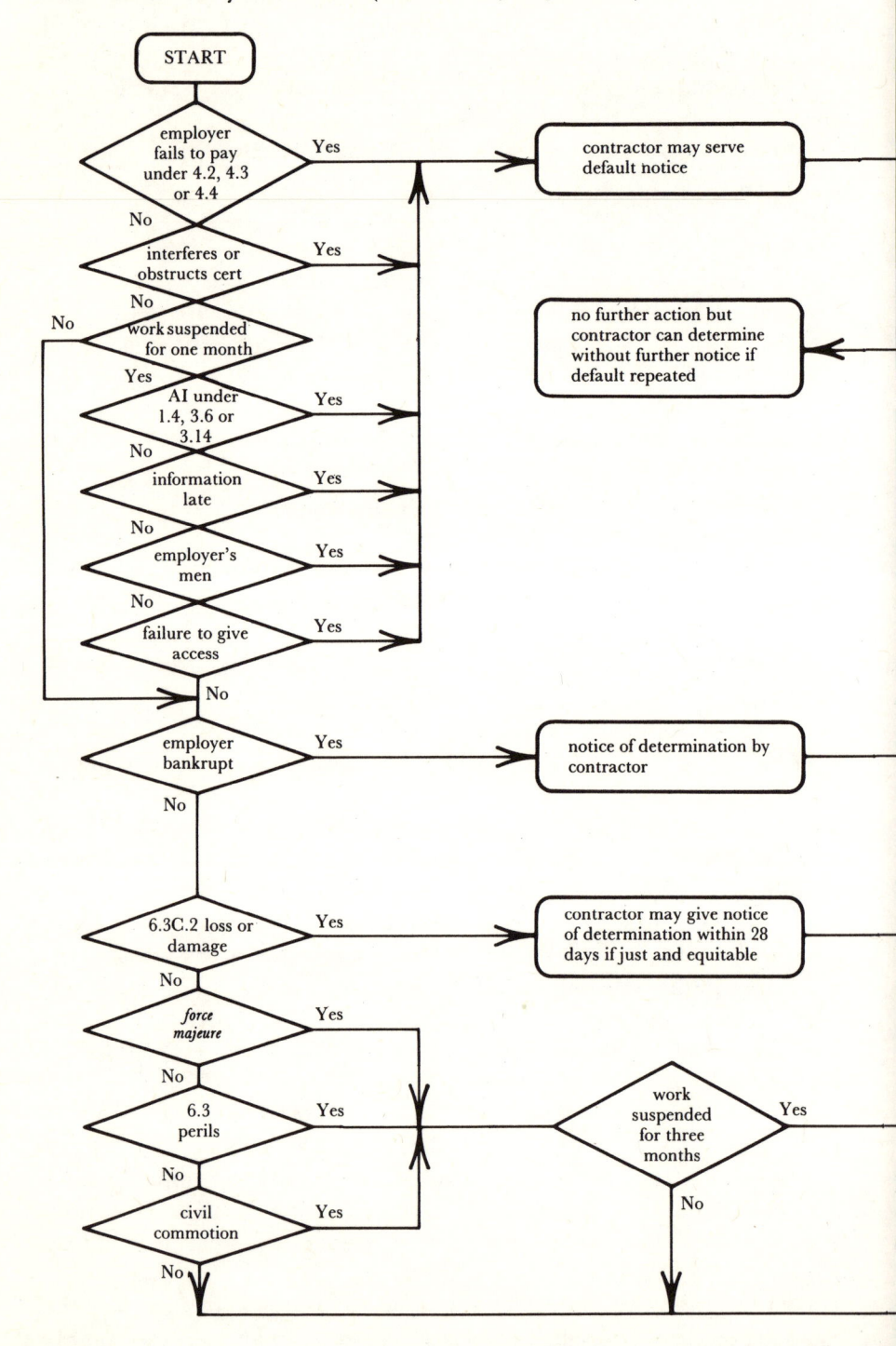

```
                    START
                      │
                      ▼
          ┌─ employer fails to pay ──Yes──▶ contractor may serve
          │  under 4.2, 4.3 or 4.4           default notice
          │        No
          │        ▼
          ├─ interferes or ──────────Yes──▶
          │  obstructs cert
          │        No
  No      ▼
  ◀── work suspended                        no further action but
      for one month                         contractor can determine
          Yes                                without further notice if
          ▼                                  default repeated
          ├─ AI under 1.4, ─────────Yes──▶
          │  3.6 or 3.14
          │        No
          │        ▼
          ├─ information late ──────Yes──▶
          │        No
          │        ▼
          ├─ employer's men ────────Yes──▶
          │        No
          │        ▼
          ├─ failure to give ───────Yes──▶
          │  access
          │        No
          ▼
          ├─ employer bankrupt ─────Yes──▶ notice of determination by
          │                                 contractor
          │        No
          │        ▼
          ├─ 6.3C.2 loss or ────────Yes──▶ contractor may give notice
          │  damage                         of determination within 28
          │        No                       days if just and equitable
          │        ▼
          ├─ force majeure ─────────Yes──▶
          │        No
          │        ▼
          ├─ 6.3 perils ────────────Yes──▶  work suspended ──Yes──▶
          │        No                        for three months
          │        ▼                              No
          ├─ civil commotion ───────Yes──▶
                   No
```

190

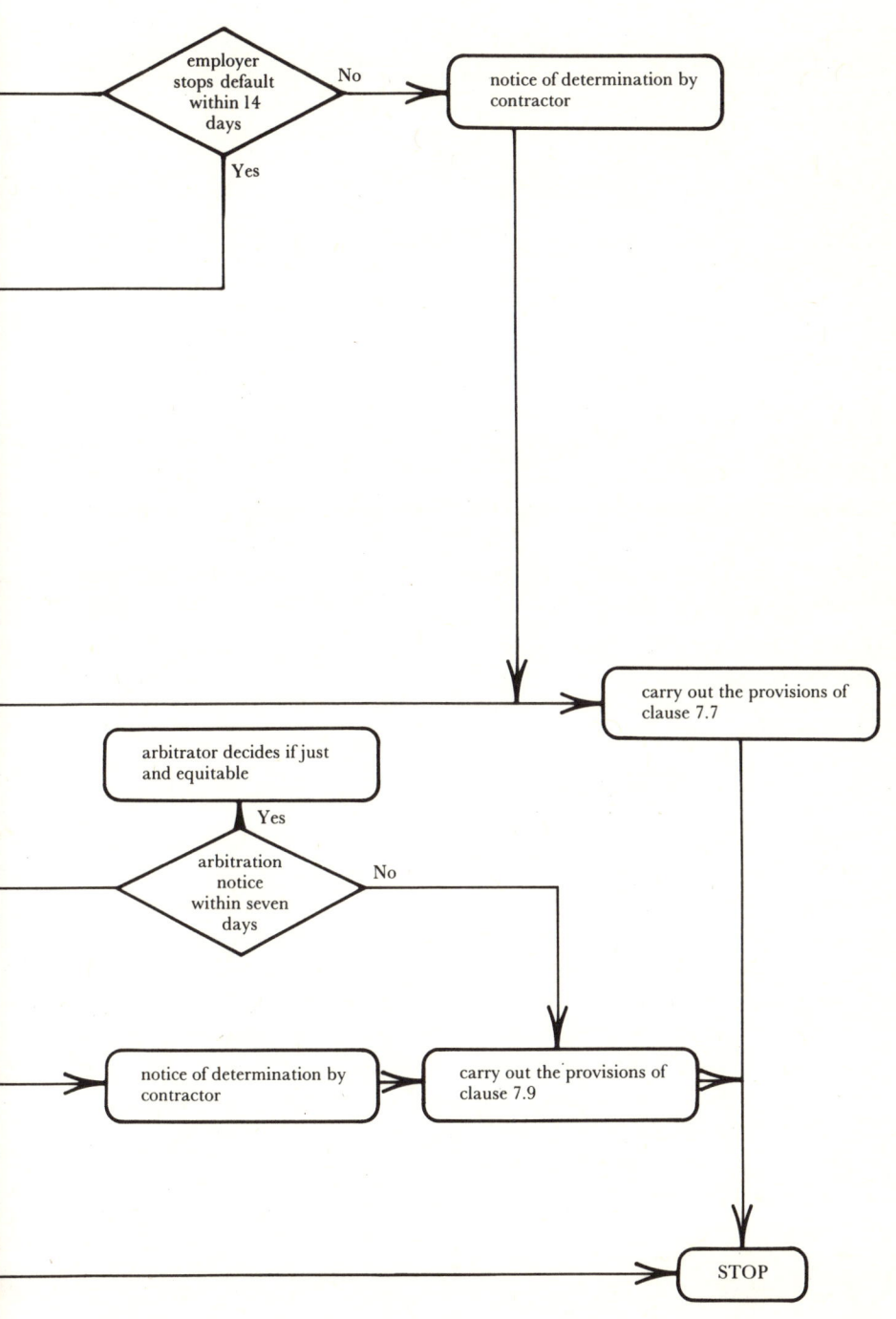

12.2 Determination by the contractor

- employer stops default within 14 days
  - No → notice of determination by contractor → carry out the provisions of clause 7.7 → STOP
  - Yes

- arbitrator decides if just and equitable
  - Yes
- arbitration notice within seven days
  - No → carry out the provisions of clause 7.9 → STOP

- notice of determination by contractor → carry out the provisions of clause 7.9 → STOP

Determination

attempting unlawful repudiation of the contract. He must send a notice to the employer (not to you) by registered post or recorded delivery, specifying one of the matters referred to in this clause. If the employer continues to make default in respect of the matter for 14 days after receipt of the notice or at any time thereafter repeats that default, the contractor may thereupon serve notice by registered post or recorded delivery to determine his employment under the contract. There is a proviso that the notice of determination must not be given unreasonably or vexatiously.

The remarks regarding postage of notices in section 12.1.2 are applicable.

The contractor has the same important power to determine his employment if the employer repeats his default or is given to the employer in clause 7.1. This means, in theory, that if the employer, after having defaulted once and been given notice, defaults a second time in payment by only one day, the contractor can serve notice of determination. In practice, such a step might be held to be unreasonable or vexatious. But there would be nothing unreasonable in the contractor's giving notice of determination if the payment on a second occasion was a week late. Most contractors are reluctant to determine because it gives them a bad reputation, whatever the reason. So in all liklihood, the contractor will send you or the employer a warning letter in the event of a second default. If he does, it will be more than you deserve.

In the hope of avoiding any chance of determination, you should consider the three grounds carefully.

The employer does not pay the contractor, etc (clause 7.5.1)

This is the employer's responsibility. You must make him aware, at the beginning of the contract, that prompt payment is vital. Your only responsibility is to ensure that the employer receives the certificate. Where possible, you should always deliver financial certificates by hand and obtain a receipt. If that is not practical, send them by recorded delivery or registered post. The period of time allotted for payment begins to run from the date of issue of the certificate, not from the date of receipt. The periods are: 14 days for interim payments and the interim payment on practical completion, and 21 days for the final certificate. Determination for failure to pay the amount shown on the final certificate would be a singularly pointless gesture; the contractor is more likely to issue a writ for summary judgment.

12.2 Determination by the contractor

The employer interferes with or obstructs the issue of any certificate (clause 7.5.2)

It is important to note that this ground refers to any certificate, not merely financial certificates. There are other certificates (Table 4.3) you are required to issue which the employer conceivably may try to prevent. It will, of course, be difficult for the contractor to prove that the employer is obstructing the issue of a certificate unless you tell him so. You have a clear duty under the contract to issue certificates. You must make it plain to the employer that he is in breach if he tries to interfere with your duty. If, despite your warning, he absolutely forbids you to issue a certificate, you are in a difficult position. Your duty is then to write and confirm the instructions given to you, but setting out the consequences to the employer (letter, Fig 12.10). You must not deliberately inform the contractor, but if he suspects and determines anyway, you will be obliged to reveal the facts in any proceedings which may follow.

The carrying out of the whole or substantially the whole of the works is suspended for a continuous period of one month, etc (clause 7.5.3)

If the carrying out of virtually the whole of the work is suspended for a month for any of the reasons set out, the contractor may determine as described. The first of the reasons is very clear. If you issue instructions regarding the correction of inconsistencies, errors in bills, etc, instructions requiring a variation, or instructions postponing the carrying out of the work, and the contractor is delayed one month thereby, he will be in very serious trouble. For any period up to one month, he would be entitled to put together a claim for loss and/or expense under clause 4.11. For any contractor handling this value of work, a one-month delay could be disastrous. This clause quite reasonably gives him the option of determination if he foresees no quick end to the suspension and he feels unable to afford to keep the site open. The clause unnecessarily emphasises that the delay must not be due to the contractor's own default.

The second reason relates to the contractor's not having received the necessary drawings and other information when required. For the delay to last a month implies a major fault on your part or a major change of mind on the part of the employer. The information must have been applied for by the contractor in writing. He must have applied for it at a reasonable time, having regard to when it would be required. Your own responsibilities under clause 1.7 do not absolve the contractor from this requirement as far as the contract is concerned, although he might have a claim at common law. A reasonable time to apply is a matter which depends on the circumstances of each individual case. The contractor

193

Determination

Fig 12.10
Architect to employer if employer obstructs issue of a certificate

Dear Sir

I confirm that a certificate under clause [*insert clause number*] of the conditions of contract is/was [*delete as appropriate*] due on [*insert date*]. I further confirm that you have instructed that I am not to issue this certificate. I am obliged to take your instructions in this matter, but you place me in a very difficult position. The contractor is certain to enquire about the certificate, and if he suspects that you have obstructed its issue, he may exercise his right to determine his employment under clause 7.5.2 of the conditions of contract. There will be serious financial repercussions for you.

In the light of the above, I look forward to hearing that you have reconsidered your position.

Yours faithfully

12.2 Determination by the contractor

has to take into account his own ordering timetable and also the length of time you will require to produce the information. It is a subject which often causes dispute and to which there is no easy answer unless you remove any necessity for requests for further information by having everything prepared at the very beginning of the contract. That is easier said than done, of course.

The third reason relates to failure or delay in work or materials which is the responsibility of the employer under clause 3.11. This clause is straightforward except for one point which could be overlooked. The contractor may determine only if there is one month's delay caused by *failure or delay* in the carrying out of the work or in the supply of the goods. If, for example, the work is carried out properly without any delay and the works are suspended as a consequence, the contractor has no power to determine. The contractor may, in such circumstances, claim an extension of time under clause 2.4.8 or 2.4.9 and loss and/or expense under clause 4.12.3 or 4.12.4.

The fourth reason is concerned with the employer's failure to give ingress or egress to the site in due time—ie, at the date for possession—and continuously thereafter until the works are completed. This is a pretty fundamental point which is the responsibility of the employer. For the clause to bite, the employer must be in possession and control of the land over which access is required. Obstructions outside the employer's land (eg, in the public highway) do not give grounds for determination under this clause, no matter how long they last. If the contract requires notice to be given by the contractor before access is granted, the period of delay will not begin to run until after you have received the notice. There is an overriding proviso, however, which could be very dangerous for the employer. You have the power to agree other arrangements for access with the contractor, and the employer will be bound by your agreement. Needless to say, you would be foolhardy to agree with the contractor anything in respect of access without the express written agreement of the employer. If you bind him to do something which proves to be impossible or expensive, you could face an action for negligence.

12.2.3 Grounds (clause 7.6)

The grounds for determination under this clause are almost exactly the same as for the contractor's insolvency under clause 7.2 (see section 12.1.3). There are two points to note. The first is that determination is not automatic, and notice must be served by registered post or recorded delivery. It is highly unlikely that the contractor would wish to continue and take his chance of being paid. The second point is that in JCT80, there is no provision for the contractor to determine on these grounds if

195

Determination

the employer is a local authority. There is no such proviso presumably because none of the events referred to is applicable to local authorities as a matter of law, though some may think that the omission is a sign of the times!

12.2.4 Grounds (clause 7.8.1)

The grounds for determination under this clause have already been covered in section 11.1.5. If the contractor wishes to determine, however, there is a proviso (clause 7.8.2) to the effect that he is not entitled to give notice if the loss or damage due to clause 6.3 perils is caused by his own negligence, the negligence of his subcontractors, or that of their respective servants or agents. This proviso is only expressly stating what must be implied—that the contractor must not be able to profit by his own default.

12.2.5 Grounds (clause 6.3C.2)

The grounds and procedure for determination under this clause are exactly the same as for the employer (see section 12.1.6).

12.2.6 Consequences (clause 7.7)

This clause lays down the procedure to be followed after determination under clauses 7.5 and 7.6. The proviso which heads this clause is somewhat different from that heading clause 7.4. The procedure is said to be 'without prejudice to the accrued rights or remedies of either party or to any liability of the classes mentioned in clause 6.1 (Injury)' arising before or during removal of temporary buildings, etc. Thus, while emphasising that accrued rights and remedies are not affected by the procedure (the employer's right to have defective work made good, for example), the contractor's liability to indemnify the employer against injury before or during the removal of temporary buildings, etc, is preserved. The procedure is as follows:

— the contractor must remove from site all his temporary buildings, plant, equipment, etc. He has a reasonable time in which to do this, but he must carry out the removal as quickly as practicable. From the time of determination until all temporary plant is removed, he must carry out his activities on site in such a way as to prevent injury, death, or damage for which he carried liability under clause 6.1. He must give his subcontractors facilities to remove their plant also. If his subcontractors default, the contractor has no duty to remove their plant, but if, after a reasonable time, some plant still remains on site, the employer would be entitled to give notice to the contractor that any

12.2 Determination by the contractor

Fig 12.11
Employer to contractor if plant left on site

Dear Sir

Clause 7.7 of the conditions of contract lays down that you must remove all temporary buildings, plant, tools, equipment, goods or materials from site with reasonable dispatch. It is now [*insert number*] days since I received your notice of determination.

Take this as notice that any of the above items remaining on site seven days after the date of this notice will be sold and the proceeds of the sale held to your credit, pending the statement of account being prepared by the architect.

Yours faithfully

Determination

plant remaining on site after, say, seven days from the date of the notice would be sold and the proceeds, after all costs have been deducted, put to the contractor's credit (letter, Fig 12.11).

— there is no express provision for the contractor to give up possession of the site, which is a pity, but the point must be academic, since if the contractor has removed all his plant, etc, he can hardly claim to be in possession.

— the contractor's liability to insure the works ends on determination, and you must immediately remind the employer to insure.

— the amount to be paid to the contractor is clearly set out. You will require the assistance of the quantity surveyor to ascertain the amounts. The contractor must be paid:

— the total value of the work done at the date of determination, less only amounts previously paid

— any sum ascertained in respect of direct loss and/or expense under clause 4.10

— the cost of materials properly ordered for the works for which the contractor has paid or is legally bound to pay (ie, because a contract has been entered into). Materials 'properly' ordered are those which it is reasonable that the contractor has ordered at the time of determination. Regard must be had to the suitability of the materials and also to the delivery period. Materials paid for become the property of the employer.

— the reasonable costs incurred by the contractor in removing his plant, etc, from site

— any direct loss and/or damage caused to the contractor by the determination. This is potentially the most damaging clause to the employer, depending on the value of the contract remaining incomplete. The contractor is to be paid the profit he would have expected to have made if the contract had run its course.

The contractor has a right to be paid, but there is no requirement for you to issue a certificate. Presumably you will simply send a statement of account to be agreed by the employer and contractor.

12.2.7 Consequences (clause 7.9 and 6.3C.2)

After either party has determined the employment of the contractor under clauses 7.8.1 or 6.3C.2, all the provisions of clause 7.7 (see section 12.2.6) apply, except that the contractor is not entitled to any direct loss and/or damage caused by the determination. This is to reflect the fact that these consequences refer to determination for causes beyond the control of the parties.

12.3 Summary

Effect of determination on other rights

The right of either party to determine is expressly stated to be without prejudice to any other rights or remedies which either party may possess. This means that all their ordinary rights at common law are preserved. Neither employer nor contractor is limited to the grounds for determination set out in the contract. If either considers that he has sufficient grounds to determine, he can rely on his common law rights.

12.3 Summary

Grounds for determination by the employer

— the contractor stops work without good reason
— the contractor fails to proceed diligently
— the contractor fails to comply with the architect's notice and serious consequences follow
— the contractor fails to comply with certain clauses
— the contractor becomes insolvent
— the contractor is corrupt and the employer is a local authority

Grounds for determination by the contractor

— the employer does not pay on time
— the employer obstructs a certificate
— the work is stopped for one month because of certain instructions, late information, delay in employer's work, or failure to give access
— the employer becomes insolvent

Grounds for determination by either party

— damage is caused to work to an existing building by clause 6.3 perils and it is fair to determine
— the work is stopped for three months because of force majeure, loss or damage by clause 6.3 perils, or civil commotion

13 Arbitration

13.1 General

Arbitration is the traditional method of settling building-contract disputes, and this tradition is continued by IFC84. Under it, disputes between the employer (or you on his behalf) and the contractor have to be submitted to arbitration (article 5). There are three exceptions:
— disputes under the fair wages provision (clause 5.7)
— disputes about Value Added Tax (supplemental condition A7)
— disputes under the statutory tax deduction scheme where statute provides for some other method of resolving the dispute (supplemental condition B8).

Arbitration, like marriage, should not be entered into lightly or inadvisedly. It is the last resort. It is costly and time-consuming, and you should do everything possible to avoid it. Some contractors will threaten arbitration over trivial matters in an attempt to persuade you to alter a decision which they dislike. You should deal with this firmly.

If the amount involved is paltry in financial terms and you are confident that your decision is correct, you may write a letter to the contractor along the lines of Fig 13.1. This often has the desired effect.

Your knowledge of the contractor is important. If it leads you to believe that he is likely to proceed to arbitration, you should adopt a different approach (letter, Fig 13.2). Firmness, coupled with tact, will often do the trick, but you must be sure of the position. Specialist legal advice should be obtained if necessary.

The majority of arbitrations are settled before the actual hearing stage is reached, but considerable preliminary costs are always incurred. It is always preferable to try to sort matters out at the meeting.

13.1 General

Fig 13.1
Architect to contractor when arbitration threatened over a small matter

Dear Sir

I have received your letter of [*insert date*] in which you say that you intend to go to arbitration unless I [*specify what contractor has required you to do. If contractor's letter was addressed to the employer, adapt appropriately*].

As I am sure you appreciate, arbitration is both costly and time-consuming, whatever the eventual outcome. In the present case, the legal and related costs are likely to outweigh any financial advantage many times over.

While I am, of course, confident that my decision would be upheld by an arbitrator, I am sure that on reflection you will consider that such a drastic course should be avoided if at all possible. In the circumstances, I hope you will agree that it would be best for us to meet as soon as possible in an attempt to solve this problem. Perhaps you will be good enough to telephone me tomorrow.

Yours faithfully

Copy: Employer

Fig 13.2
Architect to contractor when threat to seek arbitration is serious

Dear Sir

Thank you for your letter of [*insert date*] informing me of your intention to go to arbitration under article 5. [*Adapt suitably if letter is addressed to the employer*].

The step you propose is a most serious one with severe cost implications for both parties. I firmly hold that my decision, which you are seeking to question, is correct.

Obviously, neither I nor the employer can prevent you from invoking the arbitration agreement, nor would we wish to do so. No doubt, however, as a reasonable person, you will wish to avoid unnecessary time and expense, and I suggest that we meet to discuss the problem as soon as possible. If you wish to have a meeting, perhaps you will telephone me to arrange a mutually convenient appointment.

Yours faithfully

Copy: Employer

13.1 General

Although article 5 provides for disputes arising under the contract to be settled by arbitration, either party can sue in the courts. If that happens, and the defendant wishes the matter to be settled by arbitration, he must apply to the court for an order staying the proceedings. Usually this is granted, as the courts are reluctant to intervene and replace their own process with the contractual machinery agreed by the parties.

Arbitration has both advantages and disadvantages, but some of its alleged advantages are more apparent than real. Its main advantages are these:

— privacy
— the technical expertise of the arbitrator
— the arbitrator is employed by the parties as a kind of private judge. He must act as they require, subject to law. This means that matters such as the timetable and venue are fixed by agreement to suit the convenience of the parties and their witnesses
— arbitration is marginally quicker than litigation—assuming that the arbitrator is not overbooked and that the disputants do not drag their heels.

The main disadvantage of arbitration is linked to one of its advantages. It is that, as the servant of the parties, the arbitrator has few powers of compulsion. A judge's orders or direction cannot be disregarded. An arbitrator can apply to the High Court to be given certain powers (s5, Arbitration Act 1979), but this is seldom done in practice. The speed and effectiveness of arbitration depend largely on the co-operation of the parties.

13.2 Appointing an arbitrator

Unlike determination under JCT80, determination of a dispute does not have to wait until completion or alleged completion of the work, or determination or alleged determination of the contractor's employment under the contract. Immediate arbitration is available in every case.

Provided that there is a 'dispute or difference concerning this contract', either party can set the machinery in motion (article 5.1.) The first step in the procedure is for one party to write to the other, asking him to concur in the appointment of an arbitrator.

Usually, it will be the contractor who does this, but if the employer wishes to refer a dispute or difference to arbitration, you should draft a suitable letter on his behalf (Fig 13.3). It is usual and desirable to suggest the names of three people, any of whom would be acceptable as arbitrator, leaving the choice to the other party.

The arbitrator must be independent, and he must be impartial. He

Fig 13.3
Employer to contractor, requesting concurrence in appointment of arbitrator

REGISTERED POST OR RECORDED DELIVERY

Dear Sir

I/we hereby give you notice that we require the undermentioned dispute(s) or difference(s) between us to be referred to arbitration in accordance with article 5 of the contract between us dated [*insert date*]. Please treat this as a request to concur in the appointment of an arbitrator under article 5.1.

The dispute(s) or difference(s) (is)are: [specify].

I/we propose the following three persons for your consideration and require your concurrence in the appointment within 14 days of the date of service of this letter, failing which I/we shall apply to the President of the Royal Institute of British Architects/Royal Institution of Chartered Surveyors for the appointment of an arbitrator under article 5.

The names and addresses of those we propose are: [*insert names and addresses*].

Yours faithfully

Copy: Architect

13.2 Appointing an arbitrator

must not have an existing relationship with you, with the employer, or with any other person involved. For example, you should not recommend one of your partners as arbitrator!

If possible, it is sensible for the employer and the contractor to agree on an arbitrator, but in the nature of things, since they are in dispute, they are often unable or unwilling to do this. Requests to concur in the appointment of an arbitrator are often ignored. To avoid deadlock, the contract provides for an arbitrator to be appointed by a third party—the president or a vice-president of the Royal Institute of British Architects or of the Royal Institution of Chartered Surveyors, one or other of whom will be specified in the appropriate appendix entry.

The RIBA and the RICS both maintain panels of suitably qualified people from whom the appointment is made. All of them charge substantial fees which, currently, range from £250 to £750 a day, depending on the complexity of the dispute and the standing and expertise of the arbitrator.

There are two main disadvantages in having an arbitrator appointed in this way. One is that, the appointment is binding on both parties and the person appointed is almost impossible to remove. The other is that the arbitrators on the RIBA and RICS panels are often very busy people, and you lose the advantage of getting a quick hearing.

There is a time limit of 14 days before the employer or the contractor can apply to the appointing body. This runs from the date of the letter requesting concurrence in the appointment. Fig 13.4 is a suitable letter. Both appointing bodies have special forms which must be completed, and there are fees to be paid with the application. Once appointed, the arbitrator will contact both parties, with a view to arranging a preliminary meeting at which the battle lines are drawn and procedural matters are settled. By this stage, it is wise to hand over the case to a firm of solicitors conversant with arbitration law and procedure, unless you are an expert in this field. The flowchart in Fig 13.5 illustrates the procedure.

13.3 An arbitrator's powers

Arbitrators have very wide powers under the Arbitration Acts 1950-1979, and this is the meaning of the reference to 'the generality of his powers' in article 5.2. The arbitrator under IFC84 is given additional powers by the agreement itself. These are as follows:

— to direct such measurements and/or valuations as may be desirable to determine the rights of the parties and to ascertain and award any sum which ought to have been certificated.

— to open up, review, and revise any certificate, opinion, decision, requirement, or notice.

Fig 13.4
Architect to body appointing arbitrators

```
The Legal Secretary
Royal Institute of British Architects
66 Portland Place
London  W1N 4AD
[or]
The Appointments Secretary (Arbitrations)
Royal Institution of Chartered Surveyors
12 Great George Street
London  SW1P 3AD


Dear Sir

I am acting as architect for [insert name of
employer] under a contract in IFC84 form,
article 5 of which makes provision for your
President to appoint an arbitrator in default
of agreement.

Please will you send me the appropriate form of
application and supporting documentation, with
a note of the current fee payable on application.


Yours faithfully
```

13.2 Appointing an arbitrator

— to determine all matters in dispute submitted to him as if no such certificate, opinion, decision, requirement, or notice has been given.

The last two powers are very important. The arbitrator can review the exercise of your discretion and in effect substitute his own opinion for yours. This is specially important as regards such matters as loss and/or expense claims and extensions of time (see Chapter 10). Oddly, the courts do not have power to open up, review, and revise certificates or opinions as they think fit, because for them to do so would be to modify the contractual obligations of the parties. Where the parties have agreed on the arbitration of a dispute the courts will not and cannot intervene to substitute their own process for the machinery agreed between the parties.

Four matters are specifically excluded from this wide power. They are:

— clause 3.5.2: where the contractor has asked you for your contractual authority for issuing an instruction and you have responded (see section 4.1.3). Unless the matter is submitted to immediate arbitration, your answer is 'deemed for all the purposes of this contract to have been empowered' by the provision you specified

— clause 4.7: the provision in clause 4.7 about the conclusive effect of the final certificate 'except for any matter which is the subject of proceedings commenced within 21 days after' its date

— supplemental conditions C4.3 and D8: agreement between the parties as to the amount of fluctuations.

Save for these exceptions, the arbitrator's powers could scarcely be wider, and this is an important safeguard for both employer and contractor. Table 13.1 summarises the arbitrator's more important powers.

13.4 Appeals to the High Court

The Arbitration Act 1979, ss1 and 2, makes provision for

— appeals to the High Court on questions of law arising out of an arbitrator's award, and for

— the determination by the High Court of questions of law which arise during the arbitration proceedings. These powers are subject to certain requirements, one precondition being that the parties have given their consent.

Article 5.3 gives the necessary consent, and its inclusion means that disputed matters can be referred to the courts. The only proviso is that the appendix entry must state that the article is to apply. If it is not stated to apply and one party wishes to refer a legal point to the courts, he will have to meet the other requirements of the act, which means in the first case obtaining the leave of the court itself, or in the second that of the arbitrator.

Arbitration

Fig 13.5
Arbitration (article 5)

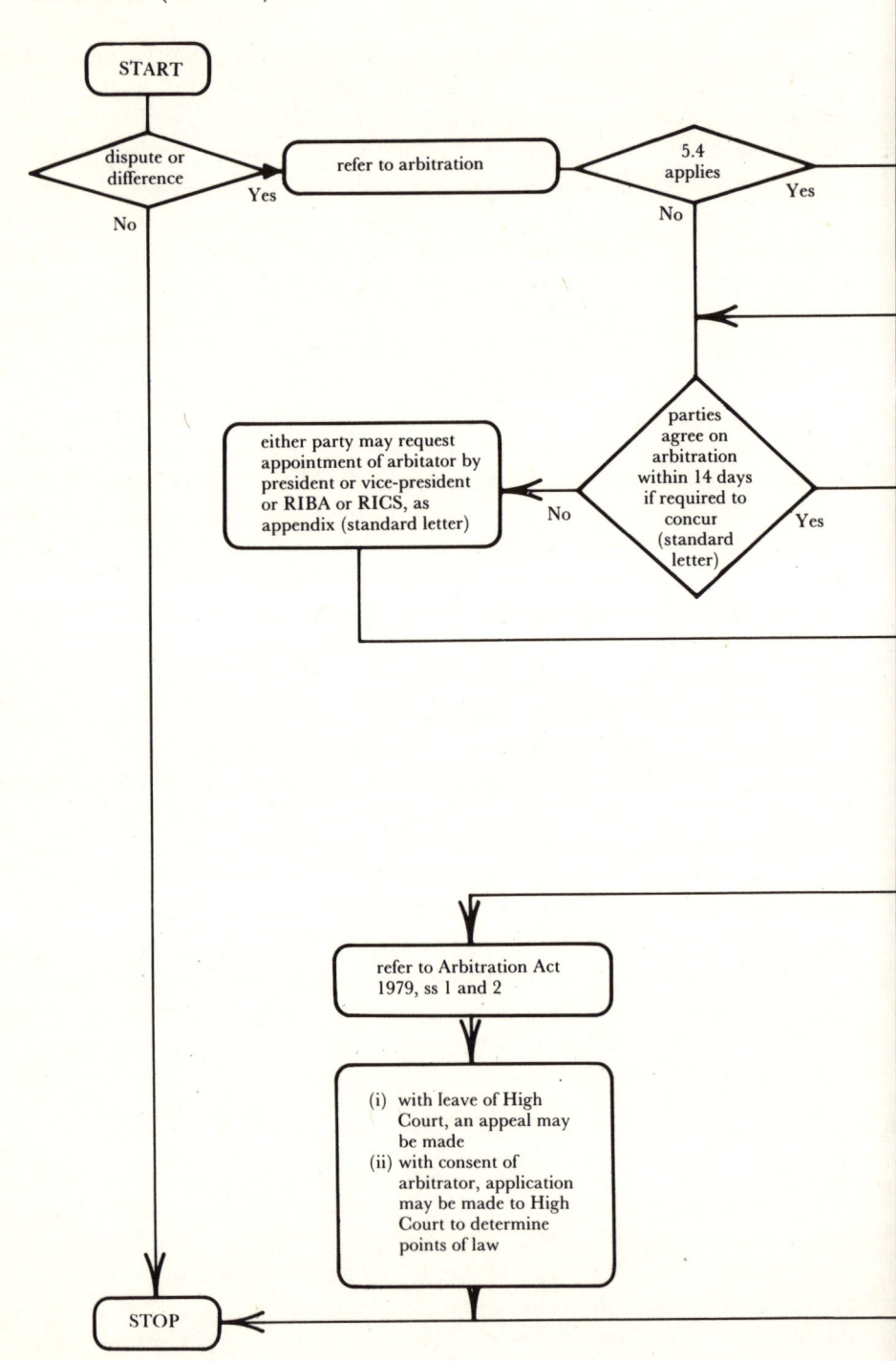

```
                    START
                      │
                      ▼
              ┌───────────────┐
              │  dispute or   │────Yes──▶ refer to arbitration ──▶ ┌─────────┐
              │  difference   │                                    │   5.4   │──Yes──▶
              └───────────────┘                                    │ applies │
                      │                                            └─────────┘
                      No                                                │
                      │                                                 No
                      │                                                 │
                      │                                                 ▼
                      │                                        ┌─────────────────┐
                      │                                        │  parties agree  │
                      │          ┌────────────────────────┐    │  on arbitration │
                      │          │ either party may       │    │  within 14 days │
                      │          │ request appointment of │    │  if required to │
                      │          │ arbitrator by          │◀No─│     concur      │──Yes──▶
                      │          │ president or vice-     │    │   (standard     │
                      │          │ president or RIBA or   │    │    letter)      │
                      │          │ RICS, as appendix      │    └─────────────────┘
                      │          │ (standard letter)      │
                      │          └────────────────────────┘
                      │                      │
                      │                      │
                      │                      ▼
                      │          ┌──────────────────────────┐
                      │          │ refer to Arbitration Act │
                      │          │    1979, ss 1 and 2      │
                      │          └──────────────────────────┘
                      │                      │
                      │                      ▼
                      │          ┌──────────────────────────────┐
                      │          │ (i)  with leave of High      │
                      │          │      Court, an appeal may    │
                      │          │      be made                 │
                      │          │ (ii) with consent of         │
                      │          │      arbitrator, application │
                      │          │      may be made to High     │
                      │          │      Court to determine      │
                      │          │      points of law           │
                      │          └──────────────────────────────┘
                      │                      │
                      ▼                      │
                    STOP ◀───────────────────┘
```

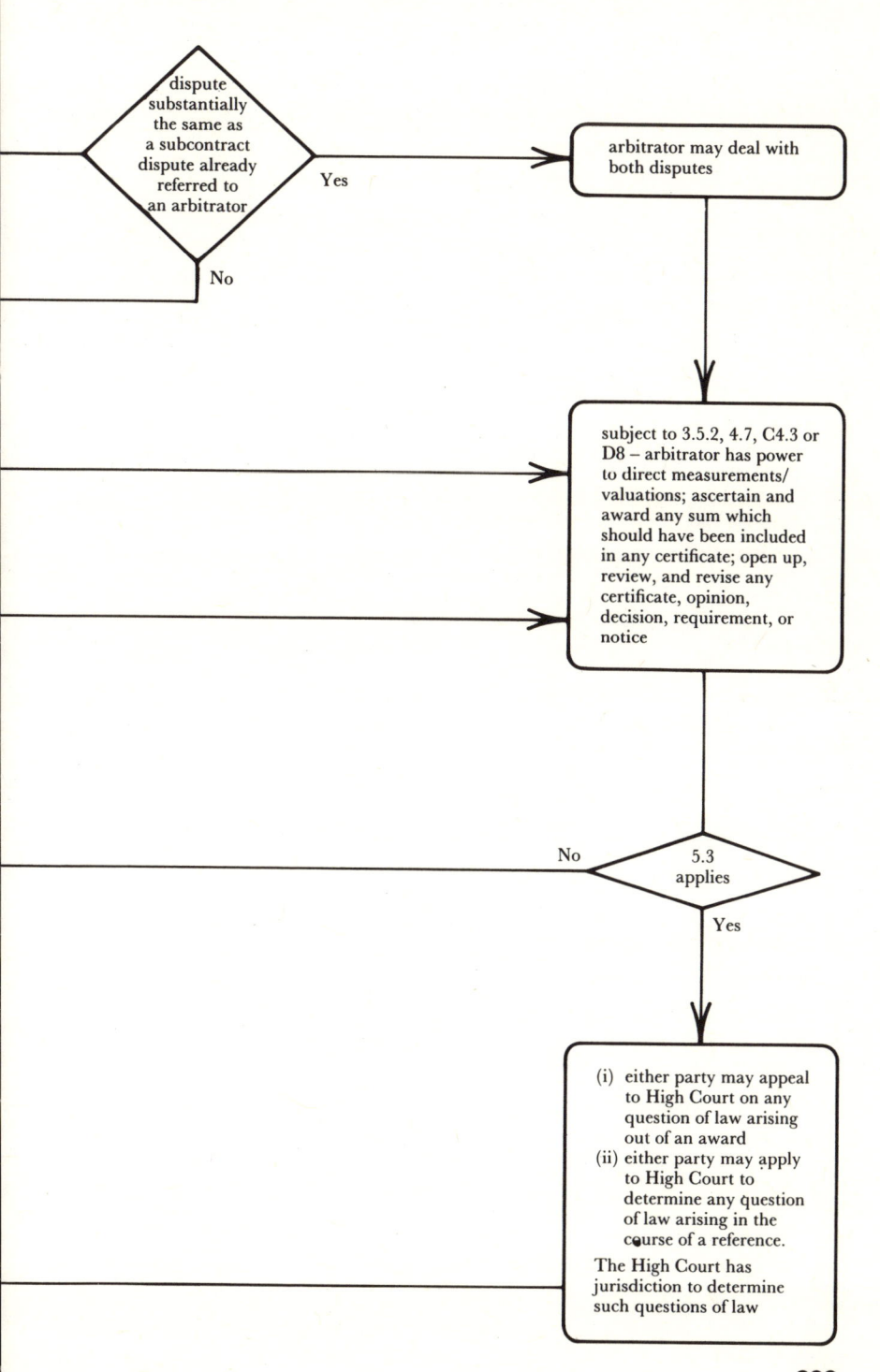

13.4 Appeals to the High Court

```
         dispute
      substantially
      the same as
      a subcontract
      dispute already         Yes  ──────────►  arbitrator may deal with
      referred to                               both disputes
      an arbitrator

          No
```

subject to 3.5.2, 4.7, C4.3 or D8 – arbitrator has power to direct measurements/valuations; ascertain and award any sum which should have been included in any certificate; open up, review, and revise any certificate, opinion, decision, requirement, or notice

```
            No  ◄──────  5.3 applies
                            Yes
```

(i) either party may appeal to High Court on any question of law arising out of an award
(ii) either party may apply to High Court to determine any question of law arising in the course of a reference.

The High Court has jurisdiction to determine such questions of law

Table 13.1
Arbitrator's powers

Power	Authority
To direct measurements and valuations to determine the rights of the parties	Article 5.2 These are the express powers conferred on the arbitrator by the parties
To ascertain and award any sum which ought to have been the subject of, or included in, any certificate	
To open up, review, and revise any certificate, opinion, decision, requirement, or notice	
To determine all matters in dispute which shall be submitted to him de novo	
To take evidence on oath	Arbitration Act 1950, s12(1)
To order disclosure and inspection of documents	1950 Act, s12(1)
To do all other things required for determining facts or law, including the power to make procedural orders	1950 Act, s12(1)
To apply to the High Court for additional and default powers if a party fails to comply with his orders	1950, s12(6) and 1979 Act, s5
To make interim and final awards	1950 Act, ss13, 14
To order specific performance where appropriate	1950 Act, s15
To award costs and to tax and settle their amount	1950 Act, s18
To award interest	1950 Act, s20 as amended

13.5 Third-party procedure

Article 5.4 is an attempt to introduce a third-party procedure into the arbitration and to give the arbitrator the same powers that the High Court has for joining parties in legal proceedings. Its legal effect is doubtful. Arbitration is a voluntary process, based on agreement, and there is no way that people can be compelled to take part in an arbitration without their consent.

There is an obvious need to join subcontractors in main-contract arbitration proceedings, and this is achieved by inserting a corresponding provision in clause 35.4 of Subcontract Conditions NAM/SC, with its reference to 'issues which are substantially the same as or connected with issues raised in a related dispute under the main contract'. A similar provision would need to be inserted in any ordinary

13.5 Third-party procedure

subcontract, and article 5.4 cannot be used if that is not done, nor could it be used to bring in consultants, such as structural engineers, without their consent. Indeed, since you are not a party to the contract, article 5.4 cannot be used to join you in the arbitration proceedings as a party. Article 5.4 applies, in any case, only where the appendix entry so states, except in the event of arbitration under clause 3.13.2 (arbitration on instructions following failure of work, etc—see section 4.1.4).

The effect of article 5.4 is to try to make provision that all the parties will join in the arbitration if the dispute referred to arbitration raises issues which are 'substantially the same as or connected with' the issues raised in a related subcontract dispute where arbitration has already been commenced. The same arbitrator is to determine all the disputes, and the article attempts to confer on him powers which he would not otherwise have.

13.6 Summary

— article 5 states that arbitration is to be used to settle disputes under the contract

— arbitration can be commenced by either party at any time

— failing agreement between employer and contractor on an arbitrator, the appointment is made by the president or a vice-president of the RIBA or the RICS

— the arbitrator has the widest powers, including power to revise any decision of the architect

— there is optional provision for legal points to be decided by the High Court if the parties wish

— arbitration or litigation should be avoided if possible.

Appendix A Form of Tender and Agreement NAM/T

The provisions for named subcontractors discussed in Chapter 8 depend for their operation on the correct and full completion of the JCT Form of Tender and Agreement NAM/T, the use of which is mandatory, whichever of the two procedures for naming subcontractors is used. The administrative burden on you is a heavy one.

The printed form opens with a warning printed in italic type: 'Should there be a separate agreement between the employer and the subcontractor relating to such matters as are referred to in clause 3.3.7 of the main contract conditions [design, etc], it should *not* be attached to the Form of Tender and Agreement, either where the form is included in the main contract documents or where it is included in an instruction of the architect . . . for the expenditure of a provisional sum'.

That separate agreement is the RIBA/CASEC Form of Employer/Specialist Agreement ESA/1, which creates a collateral contract between the employer and the named person in respect of design liabilities and associated matters. ESA/1 is summarised in Appendix C.

The guidance given in Practice Note IN/1 about the use of NAM/T is not specially helpful and the form is much more complicated than appears at first sight. It is divided into the following three parts.

Section I—Invitation to tender

The whole of this part is to be completed by you. It gives the proposed subcontractor the necessary basic information on which to base his tender. You must take great care in its completion because, in this section (and section II, below) are to be found those 'particulars' which may prevent the contractor from actually entering into a subcontract

Section I—Invitation to tender

with the proposed named subcontractor and are also referred to in the recitals of the main contract.

Section I sets out the following information:

— the name and address of the tenderer, and an invitation to him to submit a tender by completing and returning the whole form to you
— the 'priced documents' which the subcontractor will be required to provide if it is decided to name him
— any other contractual documents (called 'the numbered documents') which form part of the basis of tender
Page 1 must then be signed and dated by you.
— details of the main contract works and location, taken from the first recital of the main contract, together with any job reference
— details of the subcontract works
— details of employer, architect, quantity surveyor, and main contractor (if appointed)
— relevant main-contract information, including any changes in the printed conditions, whether the main contract is to be entered into by hand or under seal, and where the contract documentation can be inspected (if the main contractor is already appointed)
— a copy of the main-contract appendix. This is specially important as, among other things, it will fix the proposed named subcontractor with notice of any liquidated-damages provision in the main contract and gives him vital information
— access, order of works, and obligations and restrictions imposed by the employer and not covered by the main-contract conditions
— there are three alternatives, respectively where the work is to be included in the main-contract documents for pricing by the main contractor, or is to be included in a provisional-sum instruction, or in an architect's instruction naming the tenderer as a replacement subcontractor
— any deviations from the main-contract appendix entries (for example, if the named person's work is of a small amount compared with the contract value as a whole, it may be necessary to limit his liability for liquidated damages in the event of delay)
— the subcontract commencement and completion dates (at this stage, an estimate only, as the adjacent note makes clear) and period of time you will require to approve shop and other drawings after submission
— the statement that the subcontractor designate will be required to tender in accordance with the Subcontract Conditions NAM/SC form of subcontract, and indication of whether the main contractor will

have a right of reasonable objection to the person named (ie, if 12(b) or 12(c) applies)
— subcontract fluctuations
— contractor's special and general attendances.

Section II—Tender by subcontractor

This takes the form of an offer addressed to the employer and the contractor, although of course any resulting subcontract will not involve the employer. It is expressed in the alternative—and the alternatives are mutually exclusive: *either*
— 'to conclude a subcontract with the contractor by completing section III within 21 days of the contractor's entering into the main contract with the employer' (where the work is to be priced by the main contractor), *or*
— to the same effect but subject to the main contractor's right of reasonable objection. Interestingly, the reference here is only to 'item 12 (b) of section 1' and there is no reference to item 12 (c)', which refers to replacement named subcontractors.

There are various options which must be deleted, as advised in the notes on completion. One of the more interesting features is the following statement: 'This Tender, subject to any extension of the period for its acceptance, is withdrawn if not accepted by the contractor within . . . (weeks) of the date of this Tender'.

Despite this statement—which indicates an intention to hold the offer open for a stated period—under the general law, the proposed subcontractor is entitled to revoke or withdraw his tender at any time before it has been accepted by the contractor. So problems are likely to arise if in fact a proposed subcontractor withdraws or if, as is possible, the offer comes to an end through the passage of time. Lapse of time would in any event kill the subcontractor's tender; the guidance note on page 12 puts you on guard as to the position and advises you to seek an extension of the period of validity if necessary.

In section II, the subcontractor can specify any special requirements which he has and also give programme information.

When this section is signed by or on behalf of the subcontractor, it is capable of being a firm offer in law. It can be accepted only by the contractor, and you are advised to study carefully the note (dd) on page 12 which requires you to countersign the tender before passing it on to the main contractor, having deleted one of the alternatives. Note that the subcontractor is not bound to accept any changes in the information and may decide to withdraw his tender if there are changes.

Section III—Articles of agreement

You will not be concerned with the completion of section III, which is in standard legal form. On completion by the contractor and subcontractor, there will be a binding subcontract between them on Conditions NAM/SC. The articles can be entered into under seal or merely by signature. If seal is the chosen method, it will be the contractor's responsibility to see that the Inland Revenue formalities are attended to, unless you are prepared to take on this chore.

Provision is made for the names of both an adjudicator and a trustee-stakeholder under Conditions NAM/SC, and it is essential that those names and addresses be inserted if the relevant subcontract provisions are to be operable.

Appendix B NAM/SC Subcontract conditions

These conditions are deemed to be incorporated in article 1.2 of the articles of agreement in section III of the JCT Standard Form of Tender and Agreement NAM/T. The provisions are briefly set out below.

1 Interpretation, definitions, etc

This clause states that the contract is to be read as a whole. The contents of Tender and Agreement NAM/T are subject to the subcontract conditions unless otherwise specifically stated. The major part of the clause is taken up by a useful set of definitions.

2 Subcontract documents

The documents are to be those referred to in article 1.1. Nothing contained in any document issued in connection with the works imposes greater obligations than those in the subcontract documents. In the event of conflict, NAM/T prevails over NAM/SC, and these documents together prevail over any other subcontract documents. The contractor must provide two copies of any further information required, and he is to issue directions to correct inconsistencies.

3 Quality and quantity of work

Where bills of quantities are included, they are to be in accordance with SMM, and the quality and quantity of work are to be as set out therein. If there are no bills of quantities but quantities are contained in the numbered documents, they will control the quality and quantity of work. Otherwise, the subcontract documents taken together will determine the quality and quantity of the work, the contract drawings prevailing. Inconsistencies are to be corrected by the contractor's directions.

4 Subcontract sum—additions or deductions

Provisions for additions or deductions in the conditions are to be taken into account in the next interim payment following ascertainment.

5 Execution of the subcontract works—directions of the contractor

The subcontractor's obligations to complete in accordance with the subcontract documents are set out. Where approval of workmanship or materials is to be to the satisfaction of the architect, they are to be so. The contractor may issue reasonable directions in writing. The architect's written instructions affecting the subcontract works are to be taken as from the contractor. Variations will not vitiate the subcontract. The subcontractor must comply forthwith unless the variation relates to 'obligations or restrictions' in regard to access, etc, when he may make reasonable objection. The contractor may issue a seven-day notice of compliance and, if this is ignored, can employ others, pay them, and deduct the money from money due to the subcontractor. The contractor may issue directions for opening up and testing, but if the work is in accordance with the subcontract, the cost must be added to the subcontract sum. If any work or materials fail, the contractor has the power to ask that similar work be opened up and tested, provided that he uses the power reasonably. Reasonableness may be tested by arbitration.

6 Subcontractor's liability under incorporated provisions of the main contract

The subcontractor is to observe and comply with all main contract provisions as far as they relate to the subcontract, and he is to indemnify the contractor against breach, act, or omission of the main contract provisions in so far as they relate to the subcontract and against any claim resulting from the subcontractor's negligence. But the subcontractor has no liability in respect of negligence by the employer, contractor, his other subcontractors, or their servants or agents.

7 Injury to persons and property—indemnity to contractor

The subcontractor is liable for, and must indemnify the contractor against, loss, etc, arising from personal injury or death in connection with the subcontract works unless caused by the negligence of the

Appendix B

contractor or of the employer. The subcontractor must generally indemnify the contractor against loss, etc, arising from damage to property caused by the carrying out of the subcontract works unless it is caused by the negligence of the contractor or employer.

8 Insurance—subcontractor

The subcontractor must maintain insurances to cover the liabilities in clauses 6 and 7. Where appropriate, the insurance must comply with any applicable legislation. Otherwise, the insurance cover must be the sum stated in NAM/T, section II, item 2.

9 Loss or damage by clause 6.3 perils to the works and materials and goods properly on site

Where the main contract provides that the contractor or employer is to insure against clause 6.3 perils, the subcontractor is not to be liable for loss, etc, nor under obligation to insure. If the contractor insures, he must pay the subcontractor the full value of any damaged subcontract works. If the employer insures, the subcontractor must give notice to the contractor of any damage. It will be ignored in computing regular payments, but the subcontractor's work in restoring damaged work will be treated as variation work and paid accordingly. If the main contractor's employment is determined under main-contract clause 6.3C.2(b), clause 31 of these conditions will apply. The subcontractor is responsible for any loss or damage to materials on site but not yet incorporated. After their incorporation, the contractor is responsible. The subcontractor must comply with any applicable provisions in any insurance policy of the contractor or employer which relate to clause 6.3 perils.

10 Policies of insurance

When reasonably required to do so, the subcontractor must produce documentary evidence (policies and premium receipts) that he has taken out the appropriate insurance. If the subcontractor fails to insure, the contractor may himself insure and deduct the premium money from money due to the subcontractor.

11 Subcontractor's responsibility for his own plant

Property of the subcontractor on site and not for incorporation is at his sole risk as regards loss or damage not caused by the negligence of the contractor.

12 Subcontractor's obligations—carrying out and completion of subcontract works, extension of subcontract period

The subcontractor must carry out and complete the works in accordance with the dates and periods stipulated in NAM/T and reasonably in accordance with progress of the works. If it is reasonably apparent that progress is or will be delayed, the subcontractor must give notice of the delay, specifying the cause. If, for certain causes, the works are likely to be delayed beyond the period fixed for completion, the contractor must make in writing a fair and reasonable extension of time as soon as he is able. The causes are contractor's default; force majeure; exceptionally adverse weather conditions; loss or damage by clause 6.3 perils; civil commotion, strike, etc; compliance with certain contractor's directions; late receipt of information; employer's workmen; employer's materials; inability to obtain labour or materials if the provision applies; inability to obtain access; statutory undertakers' work; deferment of possession; or valid suspension by the subcontractor under clause 19.6. There is provision for extension of time if certain defaults or events occur after the expiry of the periods for completion. At any time, the contractor may make an extension of time as a result of a review of previous decisions or otherwise, provided that previous extensions are not reduced. The subcontractor must use his best endeavours to prevent delay and do all reasonably required by the contractor to proceed with the works. He must also provide such further information as the contractor requires.

13 Failure of subcontractor to complete on time

If the subcontractor fails to complete on time, the contractor must give him a notice to that effect within a reasonable time. If the contractor subsequently grants a further extension, the notice is deemed to be cancelled.

14 Disturbance of regular progress

The agreed amount must be added to the subcontract sum if the subcontractor makes a written application to the contractor within a reasonable time of its becoming apparent that he is likely to incur direct loss and/or expense because the subcontract works are being affected by any of the following: late receipt of information, opening up or testing of work found to be in accordance with the main or subcontracts, employer's workmen, employer's materials, inability to obtain access, certain architect's instructions and certain contractor's directions,

Appendix B

deferment of possession, or valid suspension by the subcontractor under clase 19.6.

If regular progress is materially affected by the subcontractor's own default and the contractor makes a written application within a reasonable time thereafter, the agreed amount of direct loss and/or expense may be deducted from monies due from the contractor to the subcontractor or it may be recoverable by the contractor as a debt. The contractor must provide such information as the subcontractor requires. These provisions are without prejudice to the other rights and remedies possessed by the parties.

15 Practical completion of subcontract works—liability for defects

If the subcontractor notifies the contractor in writing when he considers that practical completion of the subcontract works has been achieved and the contractor does not dissent within 14 days, practical completion is deemed to have taken place on the date notified. The contractor must give reasons if he dissents, and the arbitrator must then decide when practical completion has occurred. If there is no decision of the arbitrator, practical completion will be when the architect certifies practical completion of the main-contract works. The subcontractor is liable to make good at no cost to the contractor defects, shrinkages, and other faults not caused by frost, occurring after practical completion, and to accept any similar liability of the contractor under the main contract in respect of subcontract works. If the architect instructs that certain defects are not to be made good but a deduction is to be made from the contract sum, the contractor will pass on the instruction and deduction to the subcontractor insofar as it affects the subcontract works.

16 Valuation of variations and provisional-sum work

The contractor and subcontractor may agree the amount to be added to, or deducted from, the subcontract sum, in connection with a variation, before the variation is carried out. Otherwise, omissions must be valued at the relevant prices in the priced document (as identified in NAM/T, section I), while work similar in character to that in the priced document is to be consistently valued, making due allowance for any changes. If the priced document is a contract sum analysis or schedule of rates without appropriate prices, or if there is no work of similar character, if the work to be valued is not added, omitted, or substituted, or if it is not reasonable to do otherwise, a fair valuation must be made. The clause states how prime cost is to be defined in relation to daywork, deals with preliminary items, and

19 Payment of subcontractor

excludes claims for loss and/or expense which can be dealt with elsewhere.

17A Value Added Tax
This is the normal VAT clause providing for recovery of tax and clarifying the application of the Finance Act 1972 or amendment or re-enactment. It also provides that the subcontract sum is to be exclusive of tax.

17B Value Added Tax—special arrangement
This is an alternative VAT clause for use where, under VAT (General) Regulations 1980, regulations 8(3) and 21, the contractor has been allowed to prepare the tax documents in substitution for the subcontractor's authenticated receipt and where the subcontractor consents to this system.

18 Tax deduction scheme
This extensive clause deals with the provisions of the Finance (No 2) Act 1975 in regard to the deduction of income tax.

19 Payment of subcontractor
Interim payments must be made at not greater than monthly intervals, the first payment becoming due one month at latest from commencement of subcontract works on site or off site as agreed. The contractor has 17 days in which to pay from the due date. The clause details which amounts are to be included and which amounts are to be subject to retention of 5 per cent (2½ per cent after practical completion). Unfixed materials for incorporation must not be removed without the contractor's consent. Provisions are included which are intended to ensure that, if the value of any materials has been included in an architect's certificate, the goods become the property of the employer or, if the contractor has paid for them, of the contractor. An important provision allows the subcontractor to suspend his work if the main contractor fails to pay as provided seven days after receipt of the subcontractor's notice.

Final payment is due not later than seven days after the architect's final certificate under the main contract. The contractor must notify the subcontractor of the amount before the due date. Payment must be made within 14 days of the due date; the contractor's entitlement to cash discount depends on prompt payment. All documents reasonably required must be sent to the contractor before or soon after practical completion of the subcontract works. The final certificate is conclusive that, where quality and standards are to be to the satisfaction of the architect, they are to his reasonable satisfaction and that effect has been given to subcontract provisions requiring adjustment of the subcontract sum. It is not conclusive if proceedings commenced before or within 10 days of the contractor's notice to the subcontractor of the amount or the date the final payment is made, whichever is first, or in case of accidental inclusion or deduction of items or of arithmetical errors.

20 Benefits under main contract

The contractor must obtain any lawful benefits of the main contract for the subcontractor, provided that the subcontractor requests and is prepared to pay any costs involved.

21 Contractor's right to set off

This clause specifies the only rights to set off under the subcontract. The contractor may deduct from amounts due to the subcontractor any amounts agreed between them as owing and any amounts awarded in litigation or arbitration in connection with the subcontract. The contractor may set off against money due to the subcontractor under the subcontract only properly quantified amounts relating to loss and/or expense actually incurred by the contractor, provided that the contractor gives notice in writing not less than 20 days before the payment becomes due. No amounts set off prejudice the respective rights of the parties in subsequent negotiations or proceedings.

22 Contractor's claims not agreed by the subcontractor—appointment of adjudicator

If the subcontractor objects to the contractor's setting off sums due, he may seek arbitration and adjudication, giving his reasons in writing to the contractor and the adjudicator, provided that he does so within 14 days of receipt by him of the contractor's notice of intending set off. The adjudicator has power to request further written information; to order that the contractor retains the amount, or that the subcontractor be paid the amount, or that the amount be deposited with a trustee-stakeholder pending arbitration; or to order any combination of the above.

The adjudicator's decision is binding, pending the results of arbitration

27 Determination of the employment of the subcontractor by the contractor

or litigation, and he must notify the parties in writing. The trustee-stakeholder must pay any interest on the sums deposited, but he is entitled to deduct his costs. The subcontractor must pay the adjudicator's fees, but the arbitrator may make such interim orders regarding the sums as he thinks fit.

23 Right of access for contractor and the architect

The contractor and the architect and their representives must be given access to subcontract work in preparation.

24 Assignment—subletting

The subcontractor is not allowed to assign the subcontract or sublet any portion of the works without the consent of the contractor.

25 Attendance

This clause sets out the specific items of attendance which the subcontractor may expect free of charge from the contractor, including those items set out in NAM/T I and II. The subcontractor's responsibilities are detailed, together with his rights in regard to erected scaffolding.

26 Contractor and subcontractor not to make wrongful use of, or interfere with, the property of the other

There are prohibitions against wrongful use of the other's equipment and against infringement of Acts of Parliament, regulations, etc. This clause is without prejudice to the parties' rights to carry out their respective statutory or contractual duties.

27 Determination of the employment of the subcontractor by the contractor

This clause is stated to be without prejudice to any rights or remedies which the contractor may possess. The contractor may determine if:
— the subcontractor suspends the whole of the work without reasonable cause;
— the subcontractor does not proceed with the subcontract works properly, without reasonable cause;
— the subcontractor neglects to comply with the contractor's notice to remove defective works and the works are materially affected thereby, or fails his obligations in respect of remedial work to defect; or

Appendix B

— the subcontractor contravenes the provisions for assignment and subletting or fair wages.

This is provided that the subcontractor has failed to rectify his default 10 days after the contractor's written notice. If the subcontractor becomes insolvent, the contractor can determine forthwith.

Detailed provisions are included regarding the procedure following determination. Items covered are the removal of equipment from the works, employment of others to carry out the subcontract works, and payment.

28 Determination of employment under the subcontract by the subcontractor

This clause is stated to be without prejudice to any other rights or remedies which the subcontractor may possess. The subcontractor may determine if the contractor suspends the whole of the works without reasonable cause, or seriously affects the subcontractor's work by failing to proceed with his own work without reasonable cause, or fails to pay as required by the subcontract.

This is provided that the contractor has failed to rectify his default 10 days after receipt of the subcontractor's written notice. If the contractor subsequently repeats the same default, the subcontractor may determine forthwith. If the subcontractor has suspended the work in accordance with the provisions in clause 19, he cannot determine until 10 days after the commencement of the suspension.

Detailed provisions are included regarding the procedure to be followed after determination. Removal of equipment from site and payment are covered. The subcontractor is entitled to recover direct loss and/or expense.

29 Determination of the contractor's employment under the main contract

This clause provides for the automatic determination of the subcontractor's employment under the subcontract if the main contractor's employment is determined under the main contract. The consequences are to be as in clause 28, except that if the contractor's employment is determined under main-contract clause 7.8, the subcontractor will not be entitled to direct loss and/or expense.

30 Fair wages

The subcontractor is to comply with the fair wages provisions of the main contract in respect of his own employees.

35 Settlement of dispute—arbitration

31 Strikes—loss or expense
If the works are affected by strikes or lockouts, neither party may make any claim on the other therefor; the contractor must try to keep the works available for the subcontractor; and the subcontractor must try to proceed with his own work. This clause does not affect any other rights of either party under the subcontract.

32 Choice of fluctuation provisions
Fluctuations are to be dealt with in accordance with clause 33 or 34 (as noted in NAM/T, section I, item 16).

33 Contribution, levy, and tax fluctuations
Under this clause, minimum fluctuations are allowed, relating to changes in the amounts of rates or taxes payable at the date of tender. The provisions are 'frozen' after the date on which the subcontractor fails to complete, provided that the extension-of-time clause is not amended and that the contractor carries out his duty to give a decision each time the subcontractor makes application. The clause itself is long and complex.

34 Formula adjustment
This clause allows what amounts to full fluctuations based on NAM/SC Formula Rules. Provision for 'freezing' is made subect to the same conditions as in clause 33. This clause also is long and complex, detailing provisions for notices, payment, and arbitration.

35 Settlement of disputes—arbitration
Disputes are to be settled by reference to arbitration. The provisions are unremarkable and include provision for appeal to the High Court on questions of law arising out of an award or arising during the course of the reference. Provision is also made to allow joining of parties to dispute as between the employer, contractor, and subcontractor.

Appendix C The RIBA/CASEC Form of Employer/Specialist Agreement ESA/1

Under IFC84, for certain defaults of a named subcontractor, the employer can suffer loss or damage for which he has no contractual remedy against the main contractor. This is because of the terms of clause 3.3.7, which exempts the main contractor from liability to the employer in respect of a named subcontractor's failure to exercise reasonable care and skill in any of the following:

— the design of the subcontract works so far as the named subcontractor has designed or will design them

— the selection of the kinds of materials and goods for the subcontract works so far as such goods, etc, have been or will be selected by the named subcontractor

— the satisfaction of any performance specification or requirement relating to the subcontract works.

Almost inevitably there will be a design or related element in a named subcontractor's work. Clause 3.3.7 reverses the general rule of law, which is that the main contractor is responsible for subcontractors' defaults of design, fabrication, or otherwise. True, if the named subcontractor carried out the design element negligently, and the employer suffered loss, the employer would have a remedy against the defaulting named subcontractor by bringing an action for negligence. The specified areas are ones where the employer needs further protection, which is also necessary if the subcontractor fails to provide information to the architect, so causing him to give late instructions to the contractor, and, as a result, the contractor has a valid claim for extension of time and extra cost.

These are the main areas which ESA/1 seeks to cover, the device

Appendix C

adopted being a direct contract between the employer and the named person, which is intended to be completed contemporaneously with the submission of a tender in Form NAM/1. ESA/1 echoes the superseded RIBA Form of Warranty or Agreement between an Employer and a (Nominated Subcontractor) and in some respects is akin to Form NSC2 (or 2a) as used in connection with JCT80 for nominated subcontractors. It was prepared not by the Joint Contracts Tribunal but by the Royal Institute of British Architects and the Committee of Associations of Specialist Engineering Contractors and is a fairly short form.

However, its simplicity is deceptive, and you will need to take care in its completion. It can be used in two ways—designated 'procedure A' and 'procedure B'—and it includes comprehensive guidance notes on its correct completion. One of those notes (no 3) suggests that the need for it 'must be considered according to the circumstances of each project and the degree of involvement of the specialist', but we feel that its use is essential in the employer's interests in all but the simplest cases.

Procedure A is to be used when sufficient information is available for you to invite a final tender from the named subcontractor. In this case, the boxed paragraphs A and AA on pages 1 and 3 will be used. Procedure B is to be used in other cases, when paragraphs B and BB will apply. The essential difference between the two procedures, apart from matters of minor detail, is that in the first, the specialist is invited 'to submit herewith a tender for the subcontract works ... and to satisfy the requirements described by or referred to in the schedule'. This is to be used when sufficient information is available for a final tender to be requested. In the second, the invitation is to submit 'an approximate estimate', where the information is insufficient for a final tender to be obtained at that stage. Procedure A is as follows:

— you must complete the relevant parts of ESA/1. The agreement can be entered into under seal, in which case no consideration is necessary, but if it is to be entered into under signature only, £10 is payable as consideration (the price for which the employer is buying the specialist's promises), and the appropriate alternative must be deleted

— the partly completed ESA/1 is sent by you to the proposed named subcontractor at the same time as you send him NAM/T

— the proposed subcontractor will complete the remaining parts of ESA/1 by filling in part AA and signing and dating the offer when he completes NAM/T. He makes a copy of the latter form, attaches it to ESA/1, and returns both documents to you

— if you decide to recommend acceptance of the specialist's offer in ESA/1, the employer must sign and date the agreement, and the agreement is attested appropriately, thus bringing about a direct contract for the purpose specified in the form itself.

Procedure B is broadly similar, the subcontractor filling in part BB and

Appendix C

signing and dating the offer, but in this case ESA/1 has some of the characteristics of a letter of intent. The intention is that the specialist will proceed with the development of his design so as to enable a final tender to be submitted, though it is likely that some of the legal tensions thrown up by letters of intent are likely to remain under ESA/1. For example, under procedure B, the proviso states that, should the subcontract not be entered into by the specialist, the employer undertakes to pay to the specialist 'the amount of any expenses reasonably and properly incurred'. This is in contrast to the optional clause 7.2 (dealing with materials and goods), where the undertaking is merely to 'pay . . . for any such materials and goods or the fabrication of components for the subcontract works' in like circumstances. 'Expenses' has a limited meaning, and the change in wording is presumably deliberate and not inadvertent. It is not clear exactly what is intended, save that under procedure B, a limited payment in respect of design costs must be intended.

The contents of ESA/1 are as follows:

— the preliminary part, which names the employer and the specialist, identifies the architect, and states that certain specified information is being made available to the specialist. This information is detailed in the appendix on the last page and consists of information provided by the employer with the form, further information to be provided at a later stage, and time requirements referred to in the form

— in a box, the employer's invitation to the specialist to submit a tender on alternative bases as described above (A or B)—a firm offer in law, or 'an approximate estimate'. The alternatives are mutually exclusive

— a schedule (described as the 'offer and agreement') which consists of seven numbered paragraphs:

> *1* The employer's requirements described by, or referred to in, ESA/1 itself, and the qualifying statement that the specialist will satisfy the employer's time requirements 'subject to the employer providing . . .. [any] further information referred to in the appendix' at the correct time.
>
> *2* The specialist's undertaking to provide any information which he is required to provide to the architect at a time which will enable you to co-ordinate and integrate the design of the subcontract works into the design for the main contract works as a whole. This suggests early rather than later provision of such information.
>
> *3* A further undertaking by the specialist to provide you with the necessary specified information either to enable you to obtain tenders for the main contract works and include details in the main contract documents, or to allow you to provide the main

Appendix C

contractor with the appropriate information when issuing your provisional sum instruction creating a named subcontractor. There is a further undertaking to provide information at such times that you can instruct the main contractor as the contract progresses and thus avoid claims for late instructions.

4 A statement that the employer is entitled to use the specialist's drawings and information for the purposes of the main contract works.

5 A threefold undertaking by the subcontractor as to the exercise of reasonable care and skill in design and related matters. This is a limited obligation and not as extensive as that which would be imposed on a subcontractor undertaking design under the general law. It equates to the ordinary professional standard of reasonable care and skill and so, by implication, excludes the more stringent requirement of suitability for intended purpose.

6 A definitions clause—the definitions are self-explanatory.

7 An optional provision in two parts, whereby the employer may require the specialist to buy goods or materials or to fabricate components for the subcontract works before the subcontract is entered into. In that event, should a subcontract not be entered into, the employer undertakes to pay the specialist for the materials or goods so purchased, or for properly fabricated components. This is said to be 'subect to any agreement to the contrary'. On payment, the goods, etc, become the employer's property. This statement will not, of course, override any retention-of-title clause in any contract of supply entered into by the specialist, nor can it in any way affect the rights of third parties.

— the specialist's formal offer to satisfy the requirements in the alternative forms AA and BB, both conditioned by the statement that the offer is withdrawn if not accepted within . . . weeks of its date

— a statement of the consideration of £10 (plus VAT) which is payable to the specialist and to be deleted if ESA/1 is entered into under seal

— space for the specialist's signature.

— the employer's acceptance. Since this is described as '(on behalf of)' the employer (sic), it may be assumed that you may sign on the employer's behalf, but it would be best to seek his express authority in writing

— alternative attestion clauses.

229

Appendix D
Clause number index

Recital	Page
1	4
2	4, 18, 19

Article	Page
1	4
2	4, 152
3	4, 41
4	4
5	4, 53, 185, 200, 202, 204, 206, 211
5.1	203, 204
5.2	205
5.3	207
5.4	53, 210, 211

Clause	Page
1	1
1.1	58, 59, 62, 161
1.2	20, 21, 50, 71
1.3	20, 104
1.4	21, 48, 71, 139, 142, 145, 189
1.5	21
1.6	22
1.7	19, 21, 22, 73, 139, 143, 193
1.8	22
1.10	43, 71, 95, 156
1.11	71, 96, 156
2	1
2.1	62, 110
2.2	1, 82, 97, 110, 111, 143, 144
2.3	43, 102, 113, 125, 127, 133, 138, 140, 179
2.4	133, 139
2.4.1	141
2.4.2	141
2.4.3	141
2.4.4	141
2.4.5	55, 100, 112, 142
2.4.6	142
2.4.7	22, 73, 97, 143
2.4.8	86, 106, 142, 143, 195
2.4.9	106, 143, 195
2.4.10	142
2.4.11	142
2.4.12	143
2.4.13	104, 142
2.4.14	97, 110
2.6	83, 84, 85, 131, 132
2.7	83, 84, 113, 125, 127
2.8	127
2.9	62, 112, 113
2.10	50, 55, 113, 118, 119, 158, 160
3	1
3.1	94
3.2	95, 101, 102, 103, 173, 180
3.2.1	95
3.2.2	95, 96, 156
3.2.2(a)	95
3.2.2(b)	95
3.2.2(c)	96
3.2.2(d)	96

Appendix D

Clause	Page
3.3	1, 87, 96, 100, 139, 142, 145, 173, 180
3.3.1	21, 50, 86, 97
3.3.1(a)	97
3.3.1(b)	97
3.3.1(c)	97
3.3.2	97
3.3.2(a)	100
3.3.2(b)	100
3.3.2(c)	102
3.3.3	50
3.3.3(a)	102, 103
3.3.3(b)	102, 103
3.3.3(c)	102, 103
3.3.4	86
3.3.5	102
3.3.6	103
3.3.7	61, 101, 226
3.3.8	103
3.4	72
3.5	87, 104
3.5.1	49, 53, 74, 89, 179
3.5.2	74, 207
3.6	50, 139, 142, 145, 189
3.6.1	51
3.6.2	21, 44, 50, 74
3.7	18, 144, 156, 160, 161, 170
3.7.1	164
3.7.2	164
3.7.4	164
3.7.5	165, 171
3.7.6	165
3.7.7	168
3.7.8	165
3.7.9	164
3.8	51, 100, 139, 142, 145
3.9	21, 51, 73, 104, 158, 160
3.10	91
3.11	86, 87, 97, 102, 103, 104, 139, 142, 143, 145, 195
3.12	52, 53, 139, 142, 157
3.13.1	52, 54, 75, 139, 142
3.13.2	53, 75, 211
3.14	19, 55
3.15	55, 100, 139, 142, 145, 189
4	2
4.1	152
4.2	88, 154, 169, 189
4.2.1	154, 155, 161
4.2.1(a)	168
4.2.1(b)	156
4.2.1(c)	43
4.2.2	154, 155, 160
4.3	88, 113, 116, 158, 189
4.4	89
4.5	113, 160
4.6	88, 121, 127, 160, 189
4.7	58, 59, 88, 161, 207
4.8	160
4.9	168
4.9(a)	157, 158, 160
4.9(b)	154
4.10	102, 157, 158, 160, 168, 169, 198
4.11	141, 143, 145, 157, 168, 193
4.11(a)	97, 110
4.12	144, 145
4.12.1	22
4.12.2	86
4.12.3	106, 195
4.12.4	106, 195
4.12.5	55, 100, 110
4.12.6	145
4.12.7	145
5	2
5.1	72, 104, 157
5.2	72, 104
5.3	72, 104
5.4	104
5.7	173, 180, 200
6	2, 105
6.1	23, 24, 27, 196
6.2	27, 187
6.2.1	24, 25, 26
6.2.2	24
6.2.3	24, 87, 89
6.2.4	27, 28, 29, 157
6.2.5	27
6.3	30, 31, 33, 58, 141, 182, 183, 196, 199
6.3A	30, 31, 113, 187
6.3A.1	30
6.3A.2	31, 89
6.3A.3	31
6.3A.4	31, 157
6.3B	23, 27, 30
6.3B.1	31
6.3B.2	157
6.3B.3	31
6.3C	23, 27, 30, 87
6.3C.1	157
6.3C.2	32, 172, 183, 188, 189, 196, 198
6.3C.3(b)	58
7	2
7.1	172, 173, 176, 177, 178, 181, 185, 186
7.1(a)	179
7.1(b)	62, 179
7.1(c)	179

231

Appendix D

Clause	Page
7.1(d)	180
7.2	180, 185, 186, 195
7.3	172, 182, 185, 186
7.4	89, 172, 177, 185, 186, 196
7.4(a)	186
7.4(b)	58
7.4(c)	185
7.4(d)	188
7.5	189, 196
7.5.1	88, 192
7.5.2	192, 194
7.5.3	55, 112, 193
7.5.3(c)	106
7.6	189, 195, 196
7.7	189, 196, 197, 198

Clause	Page
7.8	62
7.8.1	172, 182, 184, 189, 196, 198
7.8.1(b)	183
7.8.2	196
7.9	172, 184, 188, 189, 198
8	2
8.3	19, 112, 142

Supplement	Page
A	2, 200
B	2, 89, 200
C	2, 168, 171, 207
D	2, 162, 169, 207
E	2

Of related interest

Construction Law Reports

Edited by Michael Furmston and Vincent Powell-Smith

Case law relating to the construction industry is continually developing, as new cases are tried or go to appeal, or previously unexamined problems receive judicial consideration. Many of these cases come before the Official Referees' court, which has come to specialise in this area. This series will report Official Referees' decisions and appeal cases arising from them, and will be of vital concern to those involved in construction law. The aim will be to publish three volumes per year.

234 × 148 mm   140 pp approx   cloth ISBN   (for volume 1) 0–85139–780–8   (for volume 2) 0–85139–781–6

Building Contract Dictionary

Vincent Powell-Smith and David Chappell

The *Building Contract Dictionary* is the ultimate reference book for all those concerned with contract administration, from architects and quantity surveyors to contractors and solicitors. It provides authoritative answers to these and many other questions which are troublesome in practice. Its clarity of style and lack of pomposity will also make it immediately welcome to the layman—to clients who want to know what their solicitors are talking about—and to students in the building industry. It defines and explains in detail not only those words and phrases which might cause difficulty in connection with building contracts, but also the concepts encountered in relation to contracts, such as 'standard of care' or 'foreseeability'.

Within the definitions, the *Dictionary* also refers constantly to relevant legal cases; these cases not only illustrate more clearly the definitions in question, but they also provide suitable quotes from the judgments themselves

234 × 148 mm   *480 pp*   *ISBN 0–85139–758–1*   *cloth*

AJ Legal Handbook
Fourth edition

Edited by Anthony Speaight and Gregory Stone

Law affects the practice of architecture in more and more ways, and to an increasing degree. This book has become a standard text for students and an essential office reference for architects and related professionals in the building industry.
The contents of the fourth edition reflect the various new developments in case law, particularly land law, the law of negligence and the law of limitations. Forms of contract such as the JCT form have been amended, and professional codes of practice have been revised.

Contents

Introduction to English law
Introduction to Scots law
English land law
Scottish land Law
Building contracts
Building contracts in Scotland
The liability of architects
Arbitration
Statutory authorities in England and Wales
Statutory authorities in Scotland
Planning law
English construction regulations
Construction regulations in Scotland
Copyright
Architects and the law of employment
Legal organisation of architects' offices
Architect's appointment
Professional conduct in England
Professional conduct in Scotland

'This is undoubtedly a comprehensive and authoritative guide to those areas of law of particular concern to architects and other professions involved in the building industry.'

Local Government Chronicle

297 × 210 mm   254 pp   ISBN 0–85139–751–4   paper

The Architect's Guide to Fee Negotiations

Ray Moxley

Competitive fee tendering is now a reality of architectural practice. Closely following the RIBA *Plan of Work*, this new book sets all the items that have to be negotiated at each stage, so that costs can be assigned to them and the work of all the members of the building team noted and priced. It thus serves as a check list, a negotiating instrument and a financial working tool.

Contents
The full form
The short form
Conditions of appointment
Recommended fees and expenses
Appendix: Costing work in the professional office
Software manual for Feemaster

'Should earn the gratitude of all architects in private practice who wish to survive in a competitive world.'

John Partridge, Chairman, ACA

210 × 297 mm   *160 pp*   *ISBN 0–946228–05–1*   *paper*

Professional Liability

Ray Cecil

Architects are now more than ever vulnerable to legal actions, which may occur long after a building has been completed, and due to inflation may involve far larger sums than the cost of the original building. The law here is complex; how to practise safely while still providing a professional service has become a major concern of the whole profession.

Ray Cecil is an architect writing for architects, to 'advise, guide and horribly warn' them. The reader is taken through various situations and is shown what the law seemed to be, and what actually happened. The author deals in depth with the problems of suitable and adequate insurance. He offers a combination of experienced professional advice and real-life example, with practical checklists of what to do to avoid or minimise trouble.

Lastly, Ray Cecil provides a vigorous discussion of the injustices of the present system and describes the changes the future should bring.

Contents
An outline of the law
The main areas of risk
Minimising the impact
Something has to change

'What the practitioner has been waiting for: a book written by an experienced practising architect and one who, by his own account, has been through the fire. It is well researched, and written in a highly readable style . . . I have no hesitation in saying that every practising architect should have a copy of this book.'

The Architects' Journal

234 × 148 mm   172 pp   ISBN 0–85139–956–8   *cloth*

Contractor's Claims

David Chappell

No two contractor's claims ever seem to be quite the same; they are a potential source of embarrassment to the architect and need to be understood if they are to be dealt with speedily, accurately and fairly.

This book gives practical guidance on ways to assess and determine a claim, and on what action to take once the decision is reached. Simple flow charts guide the reader through all the principal procedures. Model letters are provided, supporting chapters discuss related issues.

Here is a desk-top companion for working architects; and indeed for contractors too, since it indicates how claims may be most effectively presented. It also points up areas of risk, and outlines the good management measures necessary to minimise the need for claims to arise at all.

Contents
What is a claim?
Roles
Contractor's duties
Evidence
Techniques for dealing with extensions of time
Techniques for dealing with loss and/or expense
Claims from sub-contractors
Liquidated damages, penalties and bonus clauses
Architect's certificates
Employer's decisions

'In a claims-conscious age this is an invaluable book for both the inexperienced and the experienced reader. It is a guide ... which deals with the subject matter in clear and concise terms.'
The Architects' Journal

210 × 148 mm   136 pp   ISBN 0–85139–778–6   *cloth*

Legal and Contractual Procedures for Architects

Bob Greenstreet

This is a lucid route-map through the legal and contractual maze of everyday architectural practice. Clear flowcharts, checklists, guides to action, and sample documents enable the reader to find essential information at a glance, without having to study lengthy screeds of text. Where more detailed study might be desired, highly selective bibliographies list the precise references to be consulted on each individual aspect of law.

There has long been a need for a desk-top (or briefcase) manual of this kind, to complement the lengthier and more academic textbooks already in existence. This guide maps out clearly all the tricky paths of architectural practice for both the novice and the experienced traveller.

Contents

The architect and the law	Contract formation
The building industry	The construction phase
The architect in practice	Completion
The design phase	Arbitration

This second edition has been revised to take account of changes in case law and rules affecting professional practice.

'An invaluable book for students and young architects embarking on practice.'

RIBA Journal

210 × 297 mm   110 pp   ISBN 0–85139–369–1   paper

Professional Indemnity Claims

N. P. G. Thomas

More and more claims are being made against architects. This book is written for the architect who is concerned both to understand the ramifications of these claims and how best to avoid their arising in the first place.

The steps in resolving a claim, either by litigation or by arbitration, are described by following a hypothetical case, and simple language is employed to explain the terms used. The author goes on to point out how professional liabilities may arise; he discusses whether they can be avoided, and explains the working of the professional indemnity insurance industry.

He also draws on his experience of claims and provides a checklist of what to avoid and where to take special precautions. This book offers valuable reading for any architect, whether student or long qualified.

Contents
Litigation in the high court
The alternatives to litigation in the high court
The architect as plaintiff
Professional indemnity insurance
The concept of liability
Common causes of liability and their avoidance

'This is a book to be read by all architects. It avoids legal jargon and tedious description ... Architects, having read this book, should then keep it as quick reference whenever storm clouds appear on the horizon. It might just help avoid getting very wet.'

Building Design

210 × 148 mm   99 pp   ISBN 0–85139–748–4   *cloth*

Report Writing for Architects

David Chappell

A large part of an architect's working time is spent in preparing reports. They represent a point of contact between architect and client, and a basis for decision-making, They must ask for a sufficiently active response. They can influence the client's opinion of the architect and the chance of further commissions; and if they confuse fact with opinion, they risk dire legal consequences.

Dr Chappell has experience as an architect in private and public practice, as contracts administrator for a building firm, and as an adviser on conservation.

He discusses when reports are needed, their aims, who is going to read them, and the use of consultants. He lists various types of report, gives advice on their presentation, and lists necessary survey equipment in an appendix.

Contents

Terms of reference	Proof of evidence
The report	Use of consultants
Content	Back-up information
Reports associated with building projects	Incorporation of special clauses
	Appendix A: Equipment for property surveys
Miscellaneous reports	
Special reports for ecclesiastical property	Appendix B: Sample worked feasibility report

'A forceful reminder that clarity and discipline in presentation are as important as the contents of your report . . . will prove extremely useful . . . especially now that, because of the question of liability, it is so important to ensure that you have covered your project to a reasonable depth.'

The Architects' Journal

210 × 148 mm   126 pp   ISBN 0–85139–966–5   cloth

Contractual Correspondence for Architects

David Chappell

This is a book for architects about what to do when things go wrong, and how to avoid common problems in following through a job. By means of a brief commentary and 130 sample letters, the author takes the reader through what sometimes seems the minefield that has to be negotiated from the time of appointment to completion of the project and beyond. A copy of this book to hand in the office will provide a ready solution when unforeseen problems arise.

Following the RIBA Plan of Work, the author provides a route-map which will help architects to avoid many of these problems altogether, and, using the sample letters provided, to deal with the rest simply and straightforwardly.

The book assumes that the JCT Standard Form of Agreement has been used, but alternative recommendations are given for the Agreement for Minor Building Works.

Contents

Inception
Feasibility
Outline proposals
Scheme design
Detail design
Production information
Bills of quantities
Tender action
Project planning
Operations on site
Completion
Feedback
Appendix 1: How to write letters
Appendix 2: How to make a decision

'Problems can and often do arise on even the most well run contracts ... every architect knows they do but is rarely given any guidance as to how he should act ... Essential boardside reading for every architect involved in building contracts, however small or large.' *Architects' Journal*

210 × 148 mm   218 pp   ISBN 0–85139–775–1   cloth